PT BOAT BASES

VOLUME 1: PACIFIC
1942-1943

FRANK J. ANDRUSS SR.
HAROLD E. "TED" WALTHER

FOREWORD BY A. P. ROSS, PHD

NIMBLE BOOKS LLC

Andruss & Walther

ISBNs:
  978-1-60888-294-6 (hard cover)

PUBLISHER'S NOTES

As always, copy-editing naval history books can be a bit of a challenge, as acronyms, obscure proper nouns, and technical terms abound. Nimble Books follows *the Chicago Manual of Style* 17th edition but makes occasional exceptions. I prepared the indexes of persons, ships, and technologies and am responsible for any errors therein.

Fred Zimmerman
Ann Arbor, Michigan, USA

PT boats were highly visible in US culture in WW II. First Lady **Eleanor Roosevelt** underway on a RON 14 78' Huckins PT Boat. Taboga, March 28, 1944. (Andy Small.)

Contents

ACKNOWLEDGMENTS (ANDRUSS)

First and foremost I would like to thank my wife Stacia, who once again has put up with my endless ramblings about the PT boats and this new book. She has been a saint over the years and my rock. Special thanks to my coauthor Harold "Ted" Walther, whose never ending work with the history of the PT boats shines through. A big thank you to Charles B. Jones, who took the time on his own to head over to the National Archives to assist us with several of these photos. Your continued help with many of my projects means a great deal. Thank you to Jerry Gilmartin who came through with a few much-needed photos, and Richard Washichek who always goes the extra mile to help my cause in making the photos look good.

Our heartfelt sorry to the family of Alyce Guthrie who for many years was the heartbeat behind PT BOATS Inc. She left this earth on June 8, 2023, and many of us will miss her. Thanks to Skip Anderson who provided us with some photos from his Dad, Lt. (jg) Ross E. Anderson Jr., and Lt. (jg) Robert Davis Helsby whose photos came in handy. Special thanks to Andy Small whose help is so appreciated, and whose vast knowledge when it comes to the Huckins PT boats is never ending. Special thanks to Robert Adams for coming through with some photos, just when it was needed.

I thank the Associated Press and Acme photos for their historical photos. Special thanks to the National Archives and Federal Records Center in College Park, Maryland as well as Nav-source Navy History. I would like to thank the Department of the Navy-Naval Historical Center in Washington, D.C. and Eugene Kirkland whose vast research over the years certainly was instrumental. Special thank you to CB Photographer R.W. Spencer, MoMM1/c Robert Hart, and MoMM2/c Gordon Goosela. Thank you to Ensign Ken Prescott for photos from his personal collection, and to the United States Government printing office, Washington. Special thank you to Time Life Magazine. Thank you to Peter Deforest who several years ago sent me some wonderful photos on a CD. Thanks to GM2/c Robert Douglas and Andy Shanahan for always listening to my questions. Thank you to the US Navy Seabee Museum in Port Hueneme, CA for a few of their photographs, from their wonderful collection. Thank you, Pacific Wrecks, for all the work you do.

Special thank you to those officers and sailors that made up the base and tender forces, as well as the men of the Seebee battalions. Their continued efforts throughout the war in the South Pacific made it possible for the little PT boats to meet the enemy as we advanced across the Pacific Ocean.

Frank J. Andruss Sr.
Feeding Hills, Mass.

ACKNOWLEDGMENTS (WALTHER)

Having a naval officer for a father and growing up in a boating atmosphere on Long Island, I have always been attracted to small boats, whether they are civilian pleasure craft, or small combatants. My love for PT boats started at an early age, when my Dad bought a 1/72-scale *PT-109* model for me at the old Navy Exchange at Mitchell Field, Garden City, L.I., NY. Many models of boats, aircraft, ships, subs, and tanks followed.

My military history research has gone in several different directions, one of the subjects that remains constant, is PT boats. From an early age I gathered lots of information on personnel, boats and bases, so much so, after enlisting in the Navy, I joined the ancestor unit of the PT boats, the Special Boat Units   eams and became a Special Warfare Combatant Craft Crewman (SWCC).

It was as a member of these units that I really got to see the inside picture, and learned to appreciate and respect even more, what the World War Two PT boat crews had to endure. There are a lot of logistical needs to operate a PT boat, imagine the requirements to operate a dozen boats in a hostile environment, where parts are scare, and new supplies take a long period of time to arrive. As an example, if parts were not on hand, it took anywhere from two weeks to a month to receive repair/replacement parts for the boats. Meaning boats would be limited in abilities or out of action completely. Bases also were a home away from home that also provided, food, medical, morale, welfare, and recreation facilities. Without the bases that were built, Motor Torpedo Boat operations would not be possible.

Over the years, I have been honored to have corresponded, with these men, many for years, they were my heroes and mentors. While they are all gone now and some are now only names in history books, others are legendary figures to this day in the United States Navy. These men told me what it was like operating PT boats before and during World War Two.

These men were V.Adm. John D. Bulkeley USN (PTC RON 1, RON 3, 7, 2(2), Capt. Robert B. Kelley USN (PTC RON 1, RON 3, RON 9), Capt. Kenneth W. Prescott USNR (RON 3(2), USS *Jamestown* AGP-3), Capt. Robert B. Green, USN (PTC RON 1, Elco SHIPSUP, RON 2), Cmdr. Henry J. "Hank" Brantingham, USN (PTC RON 1, RON 3, RON 9), Cmdr. John M. Searles, USNR (Pre-War RON 2, RON 3(2), RON 31), Lt. Cmdr. Thomas Kendall USNR (RON 3(2), Lt. Cmdr. James C. Mountcastle USNR (RON 3(2), RON 35) and lastly, a fine southern gentleman, Lt. Cmdr. John L. lles, Jr., USNR (RON 3(2), RON 6, RON 5, RON 29, RON 4).

I would also thank my late father Lt. Cmdr. Harold E. Walther Sr. (USNR) who gave me my love of the Navy at day one. My Mom, who was always there and survived and thrived as Navy wife, Navy mom, and Navy grandmother.

To the love of my life, Lydia, thank you for always being there, through my Navy career, deployments and more, as a Navy wife and Navy mom. Also thanks for enduring all my historical research, *Te Amo Para Siempre.*

Harold E. "Ted" Walther Jr.
Virginia Beach, Virginia

Photo Credits

Acme Photos
Andruss, Frank J. Sr.
Anderson, Ross E. Jr., Lt. (jg)
Associated Press
Department of the Navy-Naval Historical Center in Washington, D.C.
Douglas, Robert, GM2/c
Dr. Grabow's (cover)
Foncannon, Tom
Ganley, Ned
Gilmartin, Jerry
Goosela, Gordon, MoMM2/c
Hart, Robert, MoMM1/c
Helsby, Robert Davis, Lt. (jg)
Jones, Charles B.
Kendall, Thomas E. "Tom" Jr., Ens.
Kirkland, Eugene
Liebenow, William F., Lt.
Lochen Family
McHenry, John C., QM1/c
National Archives
Navsource
Prescott, Kenneth W., Ens.
PT Boats Inc.
Small, Andy
Spencer, R. W., CB Photographer
Time Life Magazine
US Government Printing Office
US Navy Seabee Museum
Walther, Ted

PREFACE

Over the past several years, many books on the WWII PT boats have made their way into print. Those of us who were interested in the history of these little boats were starving for as much information as we could gather. We had learned just about as much as we could about the saga of *PT-109*, but we wanted much more.

I have always been a firm believer that books which provided many photographs told a story. If you're seeing text about a smoke generator or a Packard marine engine it made it that much better to see those photos. I had long thought about doing another book, but my thoughts wandered as to the subject matter. I didn't want to put out a book that had already been done by another author. I thought about it long and hard, even bouncing some ideas off my PT Boat circle of friends.

It came to me one night while watching a WWII Navy show about the advances made in the Pacific. It was then that it was decided to do a book about the PT boat bases that supported these wonderful little boats. Certainly a huge task, but one that I felt Historically needed to be told. I guess the first thing that came to mind was where to begin. The number of bases that were established was vast and reached the Pacific, Mediterranean, and Alaska. There were plenty of advanced bases, some which were nothing more than tents and shacks carved out into the Jungles of the Pacific, comfortable buildings and hotels in the Mediterranean, with bases in Alaska that were cold with terrible weather. It was soon apparent that it would be an impossible task to include all these areas into one book.

After much thought, it was decided that this first volume would encompass the advanced bases in the Pacific. For this task there was no doubt that it was going to take plenty of research and photographs. I wanted to have a coauthor on this project. Enter my friend Ted Walther, who is a walking encyclopedia when it comes to the history of these little boats. Together, we have provided a manuscript that tells the story of the struggles that the young men at these bases had to endure, while working and fighting the Japanese navy.

The hardest part of this project was deciding which bases to include. We had to be certain that we had photographs that matched the sections we were writing about. Some of the bases had virtually no photographs while others had a good amount of photographic evidence. As the reader, you may find some photos not as clear as one would like, but these are how they were taken many years ago. We chose the best possible photos we could use for the book. You may notice that some sections do not contain many photos, but this is simply because there were not many available, and all areas did not produce photos of everyday life. I hope that Ted and I have provided enough to satisfy the reader.

One must take into account the conditions that these boat crews and base force personnel had to endure. The weather of the South Pacific produced some terrible heat, humidity, and rain. Crews had to work to keep the boats in shape, all the while doing it in tough conditions, and then with not much sleep, head into the night to do battle on the ocean against armed and often superior foes. The base forces worked in equally terrible conditions, operating out of unfinished shacks and tents in undeveloped and isolated tropical harbors. It is our honor that this book represents these men.

Foreword

"A sound logistics plan is the foundation upon which a war operation should be based. If the necessary minimum of logistics support cannot be given to the combatant forces involved, the operation may fail, or at best be only partially successful."
--Admiral Raymond A. Spruance

PT boats were relatively fragile craft requiring a great deal of maintenance and large volumes of fuel to successfully operate. Not intended for prolonged operations, their crew accommodations were cramped, and the galley was limited. Throughout the war, particularly in the Pacific, the boats were constantly on the move from island to island, most of which were sparsely populated and lacking the necessary infrastructure to adequately support them. While there were a small number of tenders available, they could not provide the extensive support required by the boats. To remedy this situation, the Navy built a series of bases to fully address the needs of the boats and their crews.

Despite the critical nature of the support provided by these bases, little has been written about their history. How were they built? Who built them? Where were they? What services did they provide? What happened to them after the war? This book, researched and written by two acknowledged authorities on PT boat operations, provides answers to these questions and enhances our understanding of how they contributed to the Allied victory in WWII.

A. P. Ross II, PhD

ABOUT A. P. ROSS II

Quartermaster First Class Albert Parker Ross (NSN: 2015960), United States Navy, was awarded the Silver Star for extraordinary heroism and distinguished service in the line of his profession while serving on PT-34, Motor Torpedo Boat Squadron 3, in the Philippine Islands. His gallant actions and dedicated devotion to duty, without regard for his own life, were in keeping with the highest traditions of military service and reflect great credit upon himself and the United States Naval Service.

His son, A. P. Ross II, is an educator and ship illustrator. He is the author of several books, including co-author with John Lambert of *Allied Coastal Forces in World War II*, volumes 1 and 2 (Annapolis: United States Naval Institute Press, 2018). His ship drawings have appeared in a number of books and journals, including Norman Friedman's design history on *U.S. Navy Small Combatants* and *Nautical Quarterly*.

Abbreviations

ABCD	Advanced Base Construction Depot
ASR	Air Sea Rescue
CB	Construction Battalion
CBD	Construction Battalion Detachment
CEC	Civil Engineer Corps
COMND 7	Commander Naval District 7
COMPASEAFRON	Commander Panama Sea Frontier
COTCLANT	Commander of Training Command Atlantic
EM	Electrician's Mate
GM	Gunner's Mate
LCT	Landing Craft, Tank
LCVP	Landing Craft, Vehicle, Personnel
LST	Landing Ship, Tank
MoMM	Motor Machinist's Mate
MTB	Motor Torpedo Boat
MTBSTC	Motor Torpedo Boat Squadron Training Center
OIC	Officer in Charge
OOD	Officer of the Day
PC	Patrol Craft
PTC	Patrol Torpedo, Subchaser
PT	Patrol Torpedo boat
QM	Quartermaster
R&R	Rest and Relaxation
RON	Squadron
SC	Subchaser
SCTC	Submarine Chaser Training Center
SWCC	Special Warfare Combatant Craft Crewman
USAAF	Army Air Forces
USN	US Navy
USMC	United States Marine Corps
USNR	US Navy Reserve
USO	United Service Organizations
VB	Bombing Squadron
VMF	Marine Fighter Squadron
VMF(N)	Marine Night Fighter Squadron

INTRODUCTION

Because the PT boats could not operate without repairs and supplies, the bases were as important as were the boats and personnel. In the Pacific many of the bases that were established had to first be captured from the enemy, and debris cleared before construction could begin.

The massive amount of work to construct these bases was usually carried out by Battalions of Construction workers, known as the Seabees. It is hard to believe the number of supplies and men that it took to construct these PT bases. For instance there were 133 officers and men that undertook the job in October 1942 to build the PT base in Tulagi, at Sesapi. They first had to build an emergency outlet channel by blasting and dredging the harbor, they constructed two floating dry docks from pontoons, a 500-man camp was set up, telephone systems were installed, and they provided several carpenter details to assist with maintenance and repair of the PT boats. Later in August of 1943, they would add more shop facilities and storage areas to permit major PT Boat overhaul.

These thirsty boats also needed plenty of gasoline to operate and it was the job of the Seabees to construct tanks that would hold the gas. For instance two 1,100-barrel tanks that would hold aviation gas were constructed at Sesapi. Tank Farms popped up in Tulagi that could hold thousands of gallons of gasoline and oil for not only the PT boats, but for the fleet as well. On these bases, facilities were needed that would include Hospitals, Huts for personnel, Warehouses and mess provisions. Many times native hard woods that were produced by the Seabees logging and mill activities, were used. It should be noted that this type of work was a very difficult operation as many times the men had to work in mud, guarding themselves against crocodiles, poisonous vines, and fungus infection. As the PT boats moved up the line, this type of construction was not always the case. Many of the forward areas that the boats operated from in the Pacific were nothing more than boats tied to trees with tents set up that would provide areas for berthing, maintenance, and food preparation for hungry boat crews and base force.

Places like Banika Island in the Russell Islands were highly favorable for projected facilities because they had deep water, protected harbors, and lack of malaria. This was a good base for PT boats. Seabees constructed Quonset huts for quarters, galley, mess halls, offices, operation building and dispensaries. It should be noted that airstrips were also constructed as these bases as well as tank farms and waterfront facilities.

In other places such as Mios Woendi work began for PT Boat Base 21 as the Seabees would construct housing, messing, and collateral facilities provided for 2000 enlisted men and 250 officers. Twelve finger piers were installed to accommodate as many as 50 PT boats, together with a torpedo change pier, small-boat landing and a large pontoon pier complete with a crane for changing Packard marine engines. It is our hope that this book provides the reader through photographs, an inside look of the bases and how they provided what was needed to keep the boats and their crews ready for combat.

Seabees from Construction Battalion Detachment (CBD) 1012 disembark from a RON 17 PT Boat, for jungle training exercise, Playa Vera Cruz, Panama 1943.

Miami Shakedown Base

In winter 1940/1941, Motor Torpedo Boat Squadrons 1 and 2 traveled south, for long distance underway training, the furthest they traveled, was to Cuba. However, they stopped off in Miami, and the boats were tied up to Bayfront Park. The word went out across town about the speedy Mosquito boats being tied up at Bayfront Park and thousands of Miami residents turned out to view the boats.

Later, in April 1941, four boats from Motor Torpedo Boat Submarine Chaser Squadron One (PTC RON [Squadron] One) returned to the same area for underway training and operational testing of underwater sonar gear. The testing was deemed unsatisfactory, but and once again large numbers of residents turned out to see the boats. These experiences would serve several purposes in the future.

As World War Two began there were only three Squadrons of PT boats. RON 1 was sent to Hawaii and was at Pearl Harbor on December 7, 1941, where they were credited with shooting down two Japanese planes and damaging several more. RON 2 was at the Brooklyn Navy Yard preparing for shipment to Panama, for defense of the Panama Canal. Then there was RON 3, which was shipped to the Philippines in late August 1941, they were the ones who proved the PT boats true value in combat ... they became the Legends.

Shakedown training was essential to a new squadron's work up. At that time the majority of personnel had only the initial PT Boat crew member qualification training at Melville and the crews needed to learn to work together as a team.

Up until then, shakedown training was based in Brooklyn performed in New York waters. Then when Motor Torpedo Boat Squadron Training Center at Melville, Rhode Island was opened in April 1942, Elco RONs performed their shakedown training there, and shuttling back and forth to the Brooklyn Navy Yard.

Melville was rapidly becoming a very busy place. Some officers and enlisted students were going through the Motor Torpedo Boat Squadron Training Center (MTBSTC) and using RON 4 training boats. Meanwhile, other squadrons were going through shakedown with their boats. But the fact that the winter months up north were not good for open water underway training, prompted the Navy to look for a better location for PT Boat shakedown.

For the Higgins boats, Melville was impractical because of the great distance from New Orleans. Higgins squadrons were performing their shakedowns on Lake Ponchartrain, New Orleans, La.

In early 1943, Commander of Training Command Atlantic (COTCLANT) RAdm. D. B. Beary, US Navy (USN), investigated the possibility of creating a shakedown detail for Higgins PT boats. In April 1943 the CO of Submarine Chaser Training Center (SCTC) Cmdr. E.F. McDaniel, submitted a proposal to have the prospective Motor Torpedo Boat (MTB) shakedown detail use existing facilities near SCTC. SCTC was founded in March 1942. The Chaser Training Center was stationed at the old Port of Miami at Pier 2, near downtown on Biscayne Boulevard. (The former SCTC is located near the current Museum Park.). The first Higgins squadron to attend Miami Shakedown was RON 17. Having been informed about the formation of this new training detail, Lt. Cmdr. Walsh, CO of MTBSTC and Elco Shakedown at Melville, also suggested to COTCLANT that Elco RON's also use these facilities, beginning with RON 29.

Miami was a perfect centralized location which was an ideal training environment. Joining with the existing school was a real plus for training.

In April 1943, the PT shakedown detail was established in addition to the SCTC at Miami, FL Lt. Comdr. Alan R. Montgomery, who had taken the second Squadron 3(2) to Guadalcanal, was first commanding officer of the new unit. Montgomery worked out an intensive 14-day training program, later to be expanded to three weeks, which soon became standard for shakedown both at Melville and Miami. Severe winter weather conditions at Melville led to the decision to conduct all shakedown operations at Miami, for Elco as well as Higgins boats, starting in December 1943. The shakedown detail also conducted a large amount of experimental work, often in conjunction with the experimental program at Melville. The two commands were entirely separate but maintained close liaison in both their training and experimental programs. The first Higgins RON ordered to attend the new training detail was RON 17, The first Elco RON ordered to attend was RON 29. Once they started reporting, the squadrons would arrive either at full strength or as they did later in the war, staggered in four boat divisions.

To accommodate the new unit, SCSC had to expand its waterfront holdings, so a request was put through Commander Naval District 7(COMND 7), to acquire all the pier space at the Miami Yacht Club, located just south Pier #3, In addition an `80'x 30' building be constructed at Pier #2 that would become the Shakedown offices. Torpedo overhaul and PT maintenance repair shops. Berthing would be off site at hotels within the local vicinity, facilities for repairs were adequate, once the request was approved, five finger piers at the yacht basin could accommodate up to thirty-six boats.

On 21 April 1943, they were open for business. By the end of 1943, the main building was completed, and it housed a Shakedown staff office, machine shop, gunnery shack, electrical shop, radio/communications shop, four additional offices for squadron gear, bunk rooms with double racks for base force personnel, a conference room where classes for squadron engineers were held. In addition, Packard sent company engineers down to teach classes at SCTC. A small building was utilized for sickbay and was staffed by RON Medical Officers and Pharmacist Mates. For more serious cases The Base Sickbay was located Port

Everglades, a considerable distance away. SCTC Admin and dispersing was used for personnel records and pay purposes. SCTC Supply Department was used for all supply requests. SCTC also handled all morale, recreation and welfare needs. Bayfront Park was used as an athletic field for physical training and recreational games.

Miami Shipbuilding provided marine railways that could handle 3 boats at a time for any underwater hull repairs.

While Lt. Cmdr. Montgomery was named Officer in Charge and senior instructor of the new detail, his additional staff consisted of:

- Lt. Donald Agnew, US Navy Reserve (USNR)-Communications Officer
- Lt. Robert L. Searles, USNR -Gunnery Officer
- Lt. (jg) Leonard Nikoloric, USNR-Hull and Engineering Officer
- Ens. J. M. Flachmann, USNR Seamanship and Navigation Instructor.

As the war progressed, these officers were reassigned to other RONs and returned to theaters of war. Other combat experienced officers were assigned as instructors.

As for Officer in Charge, follow on OIC's were:
- Cmdr. Barry K. Adkins, USN
- Cmdr. Robert B. Kelly, USN
- Cmdr. Russell H. Smith, USN
- Cmdr. John Harllee, USN
- Lt. Robert W. Orrell, USN

(The last OIC secured the detail in December 1945. At that time the staff consisted of seven officers and forty-one enlisted men.).

One of the more interesting aspects taken on by the shakedown detail was experimentation of all sorts of equipment pertaining to the boats and their operations in the combat zones. This work was often performed in association with the experimental program going on at Melville. The two commands were entirely separate, but always maintained close liaison in both their training and experimental programs. Different weapons set ups were experimented with, and many ended up, after testing and recommended to the RONs coming through for training, being successfully used in combat.

One point of experimentation was the operational testing of the 70' Higgins Hellcat. A completely new design produced by Higgins Industries on the initiative of the owner Andrew Jackson Higgins.

Miami Shakedown sent Lt. Robert L. Searles USNR to witness the builders' tests on Lake Ponchartrain, and later he and fellow officers put the boat through all types of underway testing. The boat was purchased by the Navy and officially designated *PT-564*, and from the reports that were submitted, everyone who drove it, loved it. It was a pure torpedo boat in every sense of the term. Speed and maneuverability were far superior to both the 80' Elco and the 78' Higgins, it could run rings around both and it's turning radius was well inside both, and it could reverse course in nine seconds, when the average 78' Higgins took twenty-two seconds. Other positive aspects were low silhouette and minimal wake at low speeds. It was also determined that it would be easier and cheaper to build.

But while all PT officers that took her out loved the boat, the role of the Motor Torpedo Boat had changed significantly, now the boats out in the war zone were performing in more of a gunboat role, instead of the torpedo boat role. This according to Bureau of Ships, required the larger boats already in production. So the design was turned down.

Another interesting item that was tested was the Elcoplane, a series of six steps fastened to the bottom and sides of a standard 80' Elco. Testing for this was done in close coordination with MTBSTC at Melville. *PT-487* a RON 4 boat was used for the initial tests, performed in Newark Bay. The underway tests were conducted on December 16, 1943. With members of the Bureau of Inspections and Surveys present, *PT-487* did a stunning fifty-five knots with light load and 53.6 knots with full war load. 10 knots faster than a brand-new boat out of the factory! If that wasn't enough maneuvering at high speed was reported as spectacular.

As RON 29 prepared to head to Miami, PTs 560-563 had Elcoplanes installed and ran from New York to Miami. During the run, the RON 29 boats proved that the Elcoplane was outstanding at high speeds. But it was found that fuel consumption was up 25 percent and lubricant oil usage was up 75 percent at cruising speed (25-30 knots). The boats would also root or bow plunge in heavy seas. Steering was made more difficult, and acceleration times increased. The planes installed on the sides of the boats warped, and the attachment brackets broke. while it seems the Elcoplane died with the tests of *PT-560-563* RON 29 Brooklyn to Miami. Information that has recently been found suggests the Navy told them to fix the problem, The speed and maneuverability were just too good to ignore. So, they moved on to the Elco SLIPPER (which was their version of trim tabs). The Elco SLIPPER, after successful testing was later installed on *PT-613-620*, of RON 42.

As previously stated, Miami was an outstanding location for training, and for many of the officers and Enlisted men assigned to the shakedown detail it was a chance to pass on lessons learned in combat to the new crews, experiment with new gear and concepts. They provided an important role in the PT Boat training pipeline.

Motor Torpedo Boat Squadron 2 (MTB RON 2) boats heading down the East Coast to Miami May 1941. These are Elco seventy-foot PTs. (Ted Walther)

Another view of the boats of Motor Torpedo Boat Squadron 2 heading down the East Coast to Miami May 1941. (Ted Walther)

PT-19 underway in Biscayne Bay Miami May 1941. (Ted Walther)

PT-10 moored at Bayfront Park Miami May1941. (Ted Walther)

PT-15, PT-10, PT-19, moored at Bayfront Park, Miami May-June 1941. The tall building in the background is The Freedom Tower, which is on Biscayne Boulevard. (Ted Walther)

MTB RON 1 and MTB RON 2 PT boats moored Bayfront Park Miami May-June 1941. (Ted Walther)

An aerial photo Motor Torpedo Boat Miami Shakedown dated May 30, 1943. The white building at the end of far-left pier is the headquarters building for the SCTC, which shared dock, office, and shop space with Miami Shakedown. Further to the left one can see a drydock, two PT boats are in it on cradles for maintenance. (National Archives via Charlie Jones.)

Officers pre-underway briefing by Lt. Cmdr. Alan Montgomery, the first OIC of Miami Shakedown, 1943. (National Archives via Charlie Jones.)

PT-245 RON 20 sunset Miami Shakedown, a gunner's mate performing pre-underway checks on twin.50s. (National Archives via Charlie Jones.)

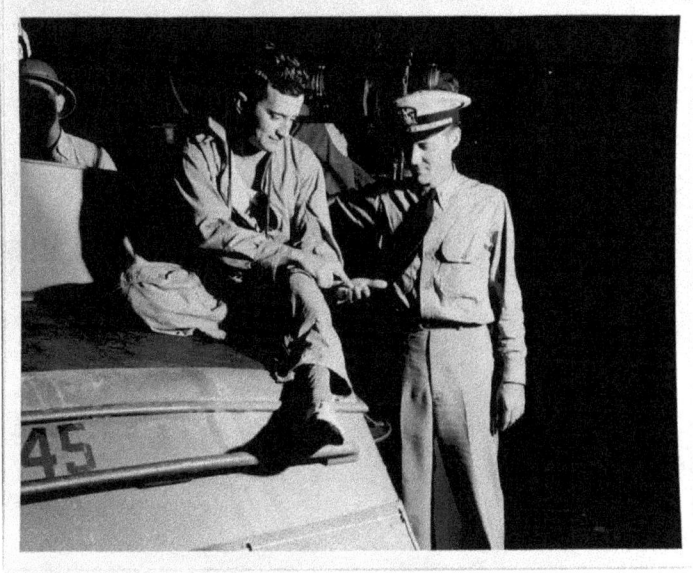

Lt. Cmdr. Alan Montgomery talks tactics with crewmember of *PT-245* RON 20. (National Archives via Charlie Jones.)

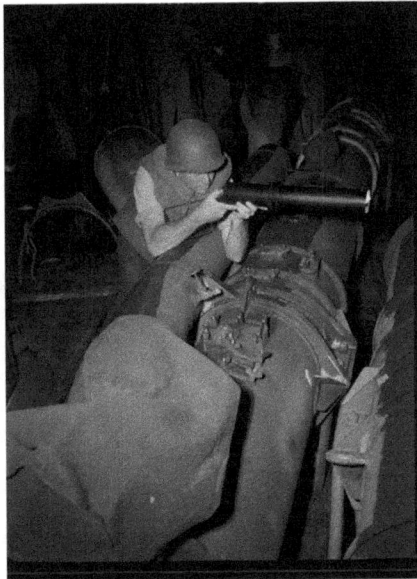

PT-245 RON 20 crewmember signaling another boat with a light signal rifle. Morse code light signals were used when radio silence was observed. (National Archives via Charlie Jones.)

Extremely rare photo of nighttime underway training. (National Archives via Charlie Jones.)

PT-245 RON 20 crewman manning the port side .50-cal turret. Nighttime underway training. (National Archives via Charlie Jones.)

RON 20 PT underway at night. (National Archives via Charlie Jones.)

RON 20 PT crossing the wake. (National Archives via Charlie Jones.)

RON 20 PT overtaking the photo boat on its port side. (National Archives photo Charlie Jones.)

A roll off Torpedo Launching Rack mounted on a boat for operational testing at Miami Shakedown. (National Archives photo Charlie Jones.)

PT-254 RON 20 testing a roll off rack Miami 1943. *PT-254* is a 78' Higgins PT boat. (National Archives photo Charlie Jones.)

PT-564 underway for operational testing, this boat was a 70' Higgins private venture. As stated in the text this was a pure torpedo boat design which all who took her out were impressed with the performance, however, it was never put in production because by then the PTs were operating more like gunboats.

PT-564 pier side, the boat was used in several different situations, here mounted forward are remote control .50 Cal machine guns.

Higgins *PT-564" The Hellcat"* Bayfront Park Miami Dock. MK-13 torpedoes mounted on roll off launching racks.

PT-261 RON 26, putting men over the side to inspect the hull, rudder, and props, possibly after having run over a sandbar. Biscayne Bay Miami 1943. (National Achieves via Andy Small.)

PT-261 RON 26 recovering hull inspection team. Biscayne Bay Miami 1943. (National Achieves via Andy Small.)

PT-304 RON 22 Miami Shakedown 1944 the Torpedoman is checking the release mechanism on the torpedo launching rack. Elcos from RON 29 in the background.

PT-308 RON 22 pier side Miami Shakedown, a crewmember is posing for the camera. The Freedom Tower is just visible just over the windshield.

PT-510 RON 35 crew Miami 1944.

PT-552 moored at Bayfront Park Miami Shakedown 1944.

PT-552 RON 29 underway offshore near Miami 1944.

PT-574 RON 38 crew poses for a group photo Miami Shakedown May 1945. (Tom Foncannon).

PT-574 RON 38 moored Miami May 1945. (Tom Foncannon).

PT-581 RON 39 underway for training Miami Shakedown Lt. (jg) Dick Secrest at wheel 1945.

PT-594 RON 40 underway for training Miami Shakedown June 1945.

RON 40 six boats underway in formation, Miami 1945.

PT-596 RON 40 coming into the pier area after training, Miami Shakedown 1945.

PT-598 RON 40 Lt. JG. T. W. Dalton, USNR CO in foreground, Miami 1945.

PT-601 RON 41 underway Biscayne Bay, Miami 1945.

PT-605 RON 41 underway during a training exercise Miami 1945.

Cmdr. Alex Michaud and Mrs. Michaud Miami Beach 1945.

The Columbus Hotel used by Navy for billeting Officers and Enlisted men.

The National Hotel also used for billeting personnel while assigned to Miami Shakedown.

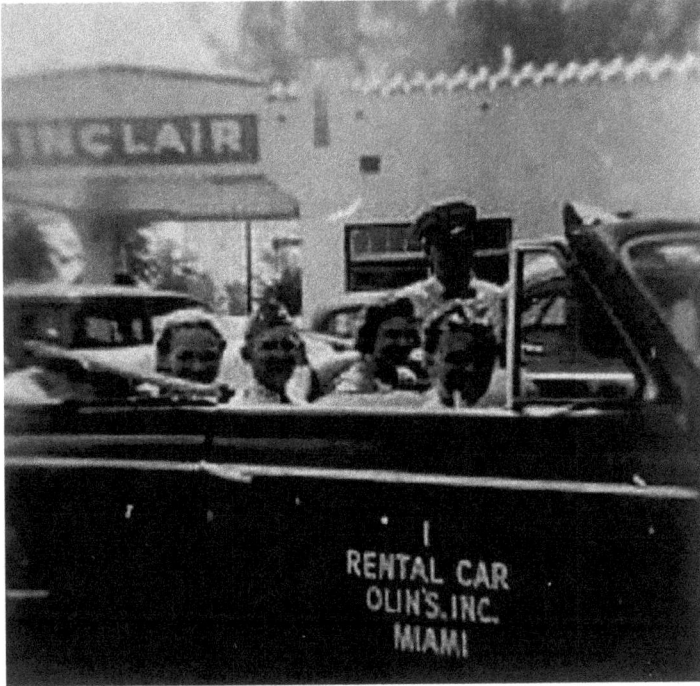

RON 38 wives in Rental car Miami 1945. (Tom Foncannon).

RON 38 officers with wives enjoying a final night together at The Flamingo, Miami 1945. From left to right: Lt. (jg) Rex "Hunk" Anderson and wife Barb, Lt. (jg) Johnnie Gordon and wife Kay, Lt. (jg) Louis Donahue, Bette and Lt. (jg) John Mitchell (later he was Attorney General of US during Nixon administration), Harriet and Lt. (jg) Gene Foncannon. (Tom Foncannon).

REFERENCES

An Administrative History of PTs in World War II. Office of Naval History. 1946. Reprinted 2013 by PageKicker Corporation, Ann Arbor.

At Close Quarters: PT Boats In The United States Navy. Captain Robert J. Bulkley Washington, DC: Government Printing Office, 1962.

PT Boat Officer-Stories From 50 Years Ago. J.E."Ted" Meredith, iUniverse.com,Inc. 2000. ISBN 978-0595006519.

Naval Station Taboga

Taboga Island has a rich history: it was at one time a liberty port for Privateer Captain Henry Morgan (a.k.a. King of The Buccaneers) and one of his lieutenants during the sacking of Panama City, Captain Robert Searle. Around 170 years ago, Taboga became the port of choice for Panama City and the mainland because the island's northern shore is well protected from the waves of the open sea and the waters are deep enough to accommodate larger ships. This is why in 1849 a British steam company serving all the American Pacific ports, The Pacific Steamship Navigation Company, chose El Morro as a main base of operations. They built large shops, facilities to ground ships and work on their hulls, a coaling station, housing for the workers, supply warehouses and a cemetery on the top of the hill. During this first year of operations, the 49'ers going to the California gold fields also preferred to stay in the healthier climate of Taboga while waiting transportation. Nevertheless, there are several graves of 49'ers who caught the disease while crossing the isthmus and died in Taboga.

Linked to the island at low tide by a sandbar, is El Morro, a small rocky islet, where at the end of the seventeenth century the Spaniards established a fort to defend Taboga. During the wars of Independence in Latin America, it was the three cannons on El Morro, manned by ten Spanish soldiers, that fought off the attacks of Englishman John Illingworth, the Chilean warship *Rosa de los Andes*, and the Peruvian Frigate *Pichincha*. Just off Taboga's northern shore, some of the little island was used as a recreation area for the PT base.

On an interesting side note, Taboga was also the site of attempted Japanese espionage when spy Yoshitaro Amana, head of a Japanese spy ring, tried to set up a commercial business on Isla Taboga so that the Japanese could ascertain what ships were transiting the Panama Canal. He was discovered in a sting operation and deported back to Japan. As well, gun emplacements at strategic points on Isla Taboga and lookout points atop El Morro made it possible to increase the security of the Panama Canal.

Already planned because of the growing world tensions and increasing Japanese aggression in the Pacific, Lt. Cmdr. Earl S, Caldwell USN Commanding Officer of Motor Torpedo Boat Squadron 2 (RON 2), was ordered to the Panama Canal Zone. On December 16, 1941, RON 2 departed with eleven Elco seventy-seven' PT boats from the Brooklyn Navy Yard, loaded aboard the aircraft ferry ships USS *Kitty Hawk* (APV-1) and *USS Hammond Sport* (APV-2), and bound for Panama to help increase defenses in the Canal Zone area and to prepare for future deployment. They arrived and were off-loaded on December 25, 1941. The boats were based out of Balboa, which had large ship repair facilities, which included large dry docks, cranes, machine shops, and more. There were also wonderful supply facilities, and Balboa was close to many of Panama Cities hot night spots for the sailor's liberty entertainment. Panama was a fantastic area for training in a tropical location, but the Balboa Harbor area was not exactly ideal for small-boat combat training and operations.

RON 2 initially divided its boats up into three groups, one group of 3-4 boats would travel to the Perlas Islands, which had several islands with deep water anchorages and were located sixty miles south/southeast from Balboa. Here the group would perform antisubmarine/anti-ship patrols, searching for Japanese Subs or shipping possibly en route to attack the Panama Canal, the group would go out and stay out for 5-7 days, on a rotating basis. Another group of boats would act as ready boats to escort any ships heading into or out of the Pacific side of the Panama Canal. The third group would be training, doing Navigation, Gunnery, or Patrol exercises.

In the meantime, The Bureau of Yards and Docks flew down from Washington to survey possible locations for a PT boat training base. They surveyed five other Central and South American locations Fonseca Bay, La Union, El Salvador, Corinto, Nicaragua, Puerto Costilla Honduras, Baltra Island in The Galapagos Islands, and Salinas, Ecuador as prospective base locations. These locations were all designated as naval base locations. It was decided that Taboga Island was ideal for this. Taboga was to become an advanced training base for PT boat squadrons operating under the Panama Sea Frontier and was set up as a war emergency project on Taboga Island, which overlooks the Pacific entrance of the Panama Canal, ten miles from the Balboa piers. The island, owned by the Republic of Panama, has a clean, sandy crescent-shaped beach, backed by a stretch of level land, rising to a series of high mounds. Once the location was approved, construction began at first with local labor. As the boats were still divided into groups, the group that was at Balboa, was further divided as needed. When construction supplies were needed two boats were used to tow Panamanian cayucas (Native canoes) and small boats out to Taboga.

Jack Searles wrote me that at times ten boats were strung out in a towline behind a PT which was just putting along, so not to swamp the smaller boats.

US Naval Station Taboga was established in August 1942, with Lt. Comdr. H. S. Cooper, USNR, as commanding officer. It was to support the PT squadron assigned as part of the naval defenses of the Panama Canal and was to serve as an operational training base for squadrons and personnel awaiting shipment to Pacific areas. It also served as a major overhaul and repair base for not only the PT's but included the Air Sea Rescue (ASR) boats that were attached to Commander Panama Sea Frontier (COMPASEAFRON).

The base was pleasantly situated on the small, mountainous island of Taboga just off the Pacific entrance to the Panama Canal. The island has a nice open harbor with sixty feet of water at low tide. The site, leased from the Panamanian Government, included a gambling casino of modern masonry construction which (though the gambling devices were removed) made a splendid recreation hall. Its purpose was to act as a main maintenance, overhaul, and operating base for a flotilla of PT boats, and as an operational training center for PT squadrons en route to combat zones. Construction began July 6, 1942, on a timber pier, two small marine railways, overhaul shops, power plant, light and power systems, refrigeration building, water storage and supply, and a radio building. Later construction included a storehouse, mess hall, barracks, quarters, A torpedo workshop, munitions storage, and numerous other facilities, services, and developments were subsequently added in quick succession.

Usable completion for several buildings was reached three weeks after work started, even in the face of lack of material, hard hand-excavating in lava soil, and slow delivery of all materials by barge from Balboa.

The Navy sent Naval Construction Battalion Detachment 1012 (CBD 1012-Seabees), under the command of Lt. (CEC) J.W. Head USNR, down from New Orleans to complete building the base on Taboga in September 1942. As soon as they arrived the work began immediately and was half done by the end of the month, when the base was commissioned, and 90 percent complete by the end of the year. CB Detachment 1012 was divided up and sent to El Salvador, Nicaragua, Honduras, Ecuador, and the Galapagos Islands, to build naval support bases at the previously mention locations.

The buildings were of frame construction on concrete foundations, many erected without specifically planned designs, time being at a premium. Later additions included two more barracks, a galley and messing facilities for 125 officers and 700 enlisted men, dry-stores building, boatswain's locker, garage, armory, berth float, pile dolphins, and a towing platform. The repair shops that were constructed were Packard and Hall Scott major overhaul and engine repair, base maintenance, machine shop, automotive school, hull repair and carpenters' shop, blacksmith shop, and torpedo shop.

A suitable water supply wasn't available, so water was brought by barge from Balboa, later evaporators were set up to augmented as an auxiliary water supply system. Sixteen 4500 lb. concrete anchors were fabricated, for mooring buoys were placed around the harbor, as mooring points for the PT's. Docking facilities were thirteen finger piers were seventy feet in length, which had 9-to-19-foot depth at mean low tide. Two marine railways that could handle 75 tons and cradles on the marine railways were changed to accommodate eighty-foot PT boats. concrete aprons for hauling boats, and a floating dry dock. When the base was first opened the boats had to refuel at Balboa. Later, on nearby Taboguilla Island, twelve large storage tanks for fuel oil and gasoline were constructed with a 1500-foot-long pier.

Taboga was the base for Squadron 2, and RON 3(2), then for Squadron 5, which relieved Squadron 2 in September 1942, and finally for Squadron

14, which in turn relieved Squadron 5 early in 1943. It was also the advanced training base for most squadrons en route to the Pacific War zone. Early Elco squadrons were shipped to Panama from New York or Norfolk. Higgins squadrons and later Elco squadrons made the run through the Caribbean to Taboga on their own bottoms.

The first squadrons conducted their own training exercises. Later, under command of Lt. Cmdr. Van L. Wanselow, USNR, the base developed an extensive training program. The base commander was designated Commander Motor Torpedo Boat Squadrons Panama Sea Frontier, and all squadrons were directed to report to him for operational control during their stay in the area.

The climate and sea conditions were ideal for all phases of advanced training. Torpedo firing, barge hunting practice, gunnery practice, navigational cruises, and joint maneuvers with aircraft could be conducted throughout the year. The many small islands in the Gulf of Panama offered unlimited opportunities for exercises in barge hunting and radar tracking.

Underway gunnery practice during day/night exercises for all caliber weapons, including rockets, was Valladolid Rock and its caves (also known locally as "The Throat"), so ricocheting fire wouldn't be an issue, the boats were ordered to approach so that all fire would be in a westward direction. Other gunnery exercises were shooting at towed target sleds, and airplane towed target socks. There was also a small arms range set up on El Morro Island. Practice Torpedo Day exercises were run on PC or SC (subchaser) patrol craft, or capital ships exiting or entering the Panama Canal sea channel. Night exercises were run against locally based destroyers. Simulated aircraft attacks were performed by locally based Army and Navy aircraft.

Thirty-One squadrons were trained in advanced tactics at Taboga. The Naval Station Taboga was decommissioned in March 1946, and all fixed improvements were turned over to the Republic of Panama.

Additional locations within Panama that were used by PT's included:

Almirante. In the summer of 1943 a small refueling base was established at Almirante, Panama, on the Caribbean side, to refuel PT boats.

Naval Station Coco Solo. Was Naval Station on the Atlantic (Caribbean) side of the Panama Canal, it was first opened in the early 1920s, and by World War Two had facilities to handle all major shipping. It was also a base for Antisubmarine Patrol Craft, Submarines, and Seaplane base. It was here that the PT's would stop after running down from Miami or New Orleans. They would refuel, and pass through the Panama Canal, en route to Taboga.

Naval Supply Depot, Balboa. Acted as the assembly point for 35 PT boat squadrons, furnishing material to complete their Squadron allowance lists, and supporting the boats with supplies and their equipment for secure stowage aboard ship. They were loaded by two 250-ton floating cranes, made available by the Panama Canal authorities.

RON 2 PT boats underway in formation headed to Naval Station Toboga Panama July 3, 1942. (Ted Walther)

PT-48 RON 2 underway headed to Naval Station Taboga Panama July 3, 1942. The officer on the foredeck looking at the camera is Ens. Thomas E. "Tom" Kendall Jr. from Minneapolis, Minnesota. He was originally a Gunnery Officer on USS Mississippi (BB-41) until he was Shanghai-ed by the Officers of RON 2. (Tom Kendall).

PT-48 Panama July 3, 1942. (Tom Kendall).

PT-60 Lt. Cmdr. Alan R "MONTY" Montgomery and Lt. John M Searles underway aboard *PT-60* 20 July 1942 Panama. (Ted Walther)

RADM Clifford Evans Van Hook, Commander 15th Naval District onboard *PT-61* Panama July 1942. (Ted Walther)

MTB Squadron Two photographed from troopship SS *President Tyler,* February 1942. (National Archives).

PT-61 en route to Taboga for commissioning ceremony followed by *PT-48* 1 AUG 1942. (Ted Walther)

PT-61 underway off Taboga *PT-47* (as *PT-1* for security purposes) following July-August 1943. (National Achieves via Charlie Jones).

PT-44 RON 2 underway with Navy Salvage divers flown down to help with the sunken S-26 (SS-131) Submarine which was accidentally rammed by USS Sturdy (PC-460), on the night of January 24, 1942. (Frank J. Andruss Sr.)

PT-47 (AS #1) underway off Taboga 1942. (Frank J. Andruss Sr.)

PT-46 Gunner Charles M. "Tubby" Kiefer underway for gunnery practice Taboga 1942. (Ted Walther)

RON 2 PTs heading down the channel toward the Sea Buoy Panama Canal and Rodman and Balboa areas in background 1942. (Ted Walther)

President Ricardo Adolfo De La Guardia (with white suit on) of Panama, at the controls of this Elco eighty-footer. The boat is underway off the coast of Panama, Rear Admiral Clifford E. Van Hook, to the President's left. Brigadier General George Brett standing behind. (Ted Walther)

R.Adm. Van Hook with Panamanian President on Elco eighty-footer, B.Gen George Brett looking at the camera. (Ted Walther)

R.Adm. Van Hook driving Panamanian President during base inspection tour of Taboga 1942. (Ted Walther)

R.Adm. Van Hook and Panamanian President on the pier at Taboga waiting prior to boarding a PT. (Ted Walther)

RON 2 and RON 3(2) boats tied up to Taboga floating pier. (Ted Walther)

A Huckins 78' PT of RON 14 leaving Balboa, Panama in 1943. Several Panamanian officers are on board for a familiarization ride. (Ted Walther)

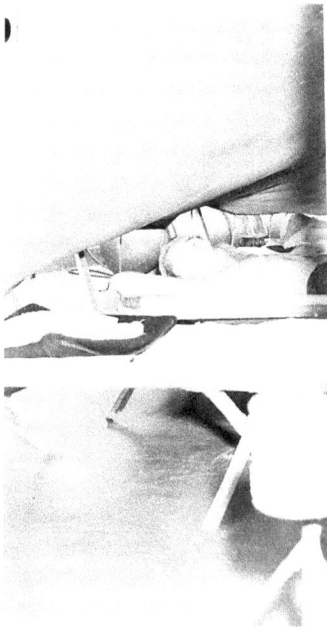

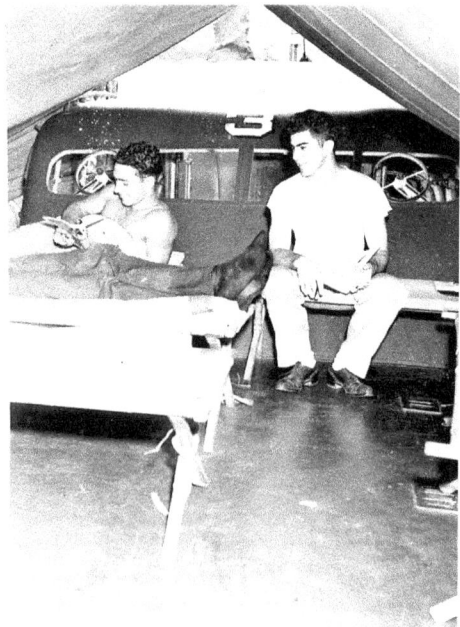

Ens. Tom Kendall, on right, in a tent erected on foredeck of *PT-48* in Taboga, Panama, July 1942. (Tom Kendall).

Lt. (jg) John M Searles below decks having breakfast with *PT-60* crew Panama July 22, 1942. (Frank J. Andruss Sr.)

RON 2 seventy-seven-foot Elco maneuvering in the harbor at Taboga 1942. (Ted Walther)

RON 5 PTs escorting USS Indiana (BB 58) into Pacific side entrance to the Panama Canal December 1942. (Ted Walther)

Gen. George Brett and R.Adm. Clifford Van Hook on board a RON 5 eighty-foot Elco in December 1942. (Ted Walther)

PT-99 of RON 14 heading to Taboga in 1943. (Andy Small)

PT-99 at Morro Island floating pier in Taboga, 1943. (Andy Small)

PT-99 of RON 14 tied up to floating pier Taboga 1943. (Andy Small)

RON 14 boat and *PT-223* and *PT-225* RON 17 practicing attack tactics. The RON 14 Huckins PT is laying a smokescreen, July 28, 1943. (Ted Walther)

PT-161 of RON 9 underway near flattop Island just north of Taboga, Panama, in 1943.
(Ted Walther)

PT-195 of RON 12 moored to the floating pier at Taboga 1943. (Ted Walther)

R.Adm. Van Hook and Under Secretary of The Navy Ralph A. Bard on *PT-195* of RON 12 Taboga 1943. (Ted Walther)

PT100 RON 14 loading a Mk VIII torpedo pierside Taboga 1943. (Andy Small).

PT100 RON 14 Going into marine railway for hull repairs. (Andy Small)

PT-100 on the marine railway, once a boat is on it, repairs can be made to the hull, shafts, props, or rudders. Once this is completed the crew would probably take advantage of the situation and scrap and repaint the hull. (Andy Small)

PT-248 RON 20 crew at Morro Island, Taboga, Panama 1943. (Ned Ganley)

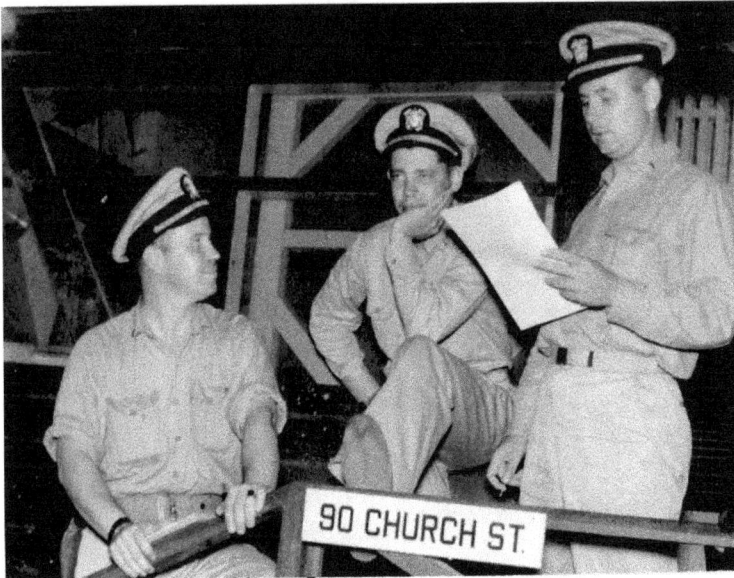

Officers in Operations Shack, Taboga, January 1943. (National Archives via Charlie Jones)

An officer coming down the 140-step staircase from the Officers' Quarters, Taboga, January 1943. (National Archives via Charlie Jones)

First Lady Eleanor Roosevelt getting some good Navy chow at the Mess Hall Taboga, March 28, 1944. (Ted Walther)

Under Secretary of the Navy Bard at the Officers' Club on Taboga. (National Archives via Charlie Jones)

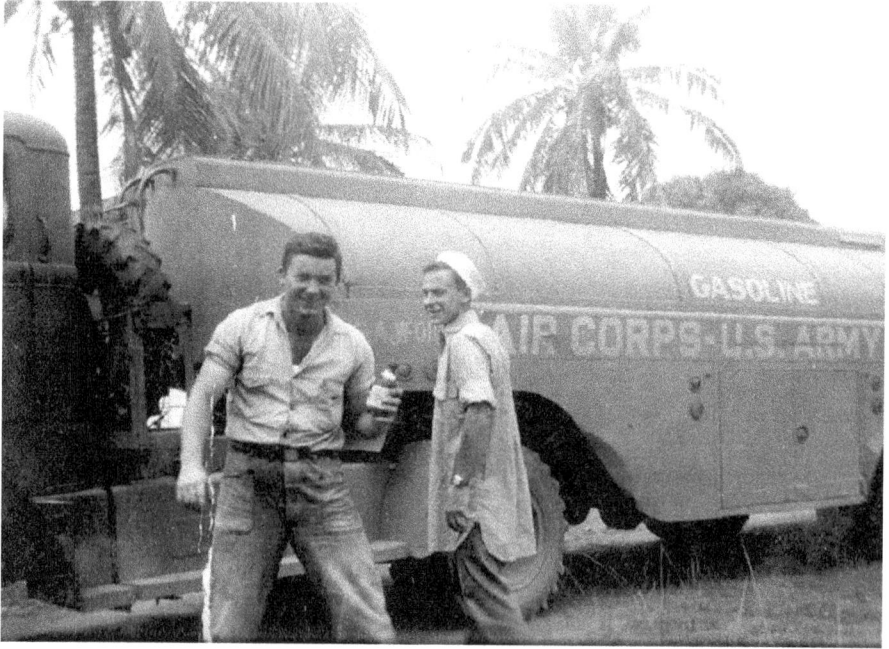

RON 14 James Lochen and fuel truck on loan from Army Airfield Albrook, Taboga 1943. (Lochen Family via Frank J. Andruss Sr.)

Aerial photo showing Naval Station Taboga, photo dated December 23, 1942. (National Archives via Charlie Jones)

Aerial photo showing Naval Station Taboga, photo dated February 1943. Notice the floating drydock in the lower right. (National Archives via Charlie Jones)

Another aerial view looking south. During high tide the sand area in the foreground is under 2-3 feet of water. This photo is dated December 23, 1942. (National Archives via Charlie Jones).

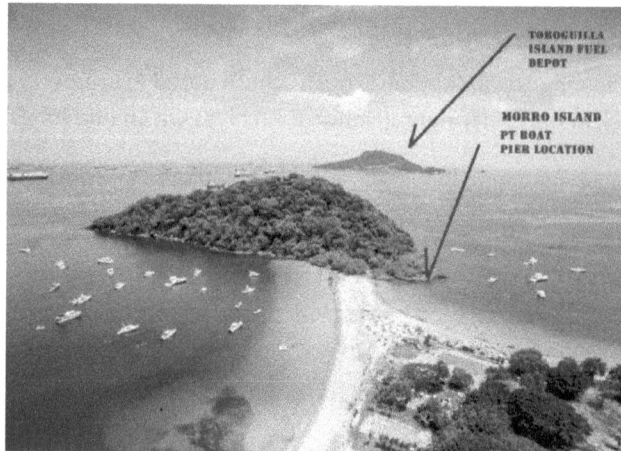

A current aerial view of the same area taken in March 2023. El Morro Island looking from above the hill on Taboga Island. Converted to grayscale. (Ted Walther)

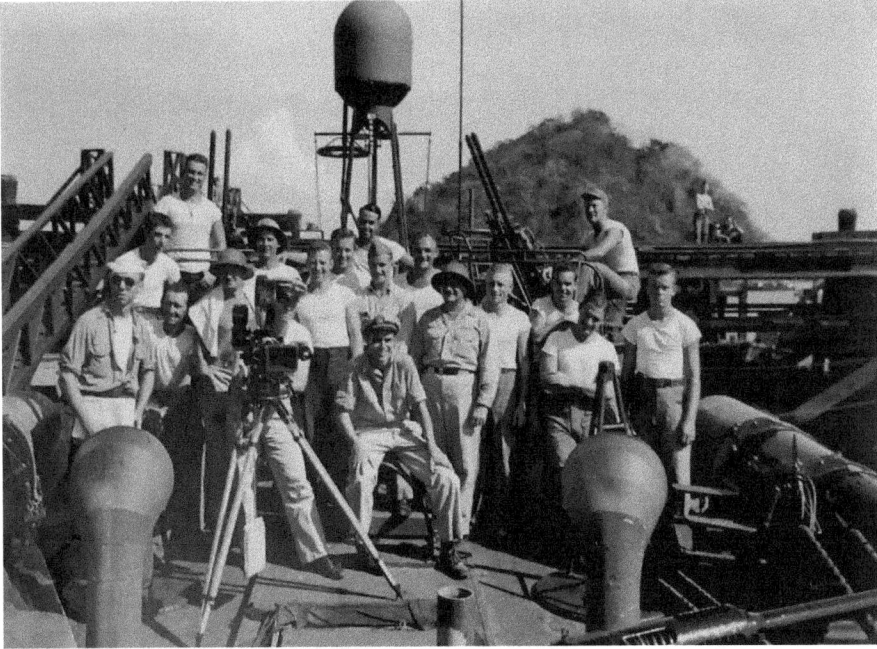

PT-483 RON 32 crew Taboga 1944. (Ted Walther)

PT-483 of RON 32 pierside Taboga 1944. (Ted Walther)

PT-578 of RON 39 moored in the harbor to an anchor buoy, Taboga 1945. (Ted Walther)

RON 32 tied up at the pier Pedro Miguel Locks, Panama 1945. The closest boat is *PT-485* "The Saint." (Ted Walther)

RON 32 PT leaving Morro Island pier 1945. (Ted Walther)

A PT from RON 32 maneuvering in Taboga Harbor 1945. (Ted Walther)

PT-479 of RON 32 underway for torpedo practice 1945 (Ted Walther).

Espiritu Santo

Six months after the bombing of Pearl Harbor, the Japanese had reached their maximum efforts in the conquest of the islands of the Pacific. They were in control of the Philippines and all the Dutch East Indies. They had gained a foothold on the north coast of New Guinea and were heading for Port Moresby on the southern shores. It was the main objective of the US forces to remove the Japanese from Guadalcanal, and that objective would begin on August 7, 1942, when the 1st Marine Division landed on the beaches of Guadalcanal and Tulagi, a fight that would last for more than a year.

When our forces had landed at Guadalcanal, we had virtually no waterfront facilities to speak of. Supplies had to be off-loaded from ships, using light landing craft, pontoon barges, and tank lighters. The 6th CB Battalion and the Marines did their own off-loading, and some timber piers were constructed. During 1943 the Seabees built finger piers, pontoon barges and other piers. All manpower was assembled to construct airfields, but after that they began to concentrate on road construction. In the last couple of months of 1942, the 14th and 26th CB Battalions built some ninety-six miles of road.

When the Japanese moved into the Solomon's and began construction of airfields on Guadalcanal, an Allied air base in an advance area became vital. The choice of Espiritu Santo, 630 miles southeast of Guadalcanal, in the New Hebrides, as a site for a major Army and Navy operating base, brought the US bombers 400 miles closer to the Japanese positions and provided a staging area for the forthcoming Allied invasion of the Solomon's. The base provided aircraft facilities capable of supporting heavy bombers, fighters, and two carrier groups; an accumulation of ammunition, provisions, stores, and equipment for offensive operations; repair and salvage facilities for all types of vessels. It became a vital link between Henderson Field on Guadalcanal and the airfields at Noumea and Efate. Espiritu Santo is the northernmost and largest of the New Hebrides Islands.

Espiritu Santo was originally established for major overhaul and engine repair. However later as the war moved Northward so quickly, hull repair was virtually eliminated, and the base was devoted to Packard marine engine overhaul and small-boat hull repair. This base was well laid out, and in 1944 was very well maintained, neat and clean. It had a compliment of 350 officers and men. Facilities were upgraded to include a concrete ramp, one 40 x 100 wood shop, excellent tools, four engine overhaul shops that were completely equipped with a water test stand, several 40 x 100 storage huts and adequate open storage space. Seabees also built storage huts with parts for the Elco and Higgins PT boats, plus parts for the Packard marine engines. Packard marine engine overhaul at this base consisted of engine tear down, complete examination, engines cleaned, and parts replaced.

By 1944, the base could effectively overhaul some 48 to 75 engines per month. This was enough to supply at least 166 boats in both the South and Southwest Pacific areas. Early in the war, Espiritu Santos was also the off-loading point for PT boats that were transported by ship. The boats were then towed by destroyers to Tulagi. Only two PT boat squadrons (RON 32 and RON 37) would be based here.

Espiritu Santos was also a staging and training area for United States Marine Corps (USMC) and USN Aviation squadrons moving up to the Solomon's. By 1944, this base was well supplied in Packard engine parts, as well as other parts for both the Elco and Higgins boats.

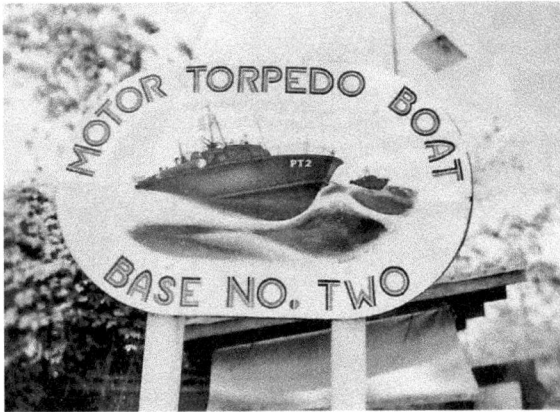

This hand painted sign greeted those that entered the base at Espiritu Santo. Primitive at first, the base over time would become much more established as Seabee units and Base force personnel would erect much-needed warehouses and other buildings that would enhance the bases engine overhaul status. (National Archives)

Early on as a base, quick set up mess areas such as this would be nothing more than a tent. Blocks with grates would serve as stoves to quickly fed hungry boat crews and base forces. The improvised mess areas were often camouflaged and compact. As time marched on better facilities would take the place of tents, making things much easier for the cooks. (National Archives)

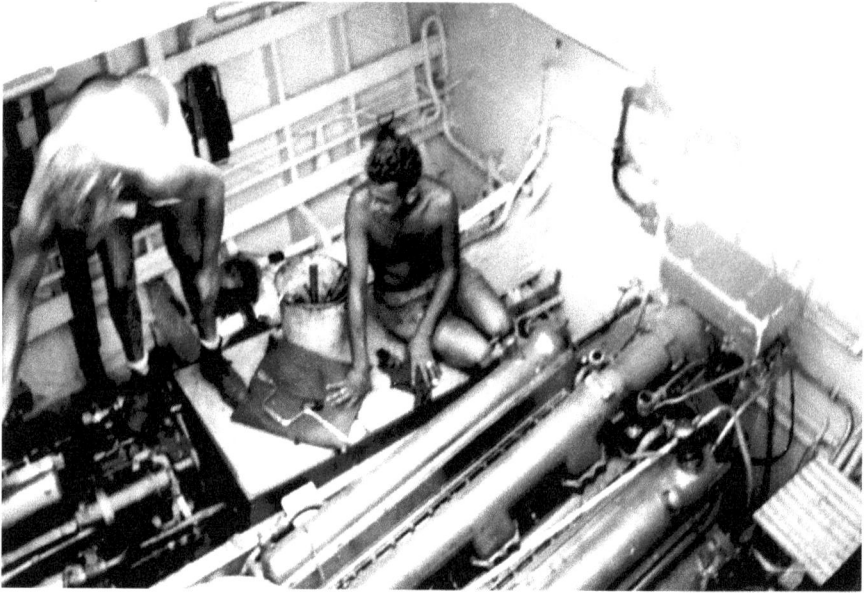

Two Motor Machinist Mates complete the engine tear down process on this boat as part of the maintenance program at the base. By 1944 the base could effectively overhaul some 48 to 75 engines per month. This was enough to supply at least 166 boats in both the South and Southwest Pacific areas. (Frank J. Andruss Sr.)

Engine overhaul at Espiritu Santo would be the main objective in keeping good running engines for the boats. Engine overhaul at this base would consist of engine tear down, complete examination, engines cleaned, and parts replaced. PT boats depended on their very life for engines that would not quit in combat. (Frank J. Andruss Sr.)

A typical Packard marine engine Test bed one would find, at several bases that had become overhaul bases. Testing the powerful 4M-2500 Packard Marine engines was critical. These testing stations of the engines would quickly check cylinder banks, pistons, crankshafts, and connecting rods. Fresh water, fuel, and oil tanks are visible here. (Frank J. Andruss Sr.)

A look at the Espiritu Santo base and the many Quonset style warehouses that would be erected here as the war went on. These areas were crucial to the base's smooth operations, giving base force personnel much-needed workspace and berthing. (PT Boats Inc.)

A look at the main gate at the base. By 1944 this base was well supplied in Packard engine parts, as well as other parts for both the Elco and Higgins PT boats. Espiritu Santos was also a staging and training area for USMC and USN Aviation squadrons moving up to the Solomon's. (National Archives)

A nice look at a portion of the base showing dock space which in the photo shows Squadron 37 PT boats. This was taken around 1945 and shows the marked improvement of the base's facilities. Early in the war, the base was also the off-loading point for PT boats that were transported by ship. The boats were towed by destroyers to Tulagi. Only two PT boat squadrons (RON 32 and RON 37) would be based here. (National Archives)

Overall aerial look at Espiritu Santo late in the war, showing the massive buildup of this base. Built at this base were personnel housing, piers, roads, shops, power plants, water plants and large storage depots with fuel, ammunition, food and other consumable supplies. (National Archives)

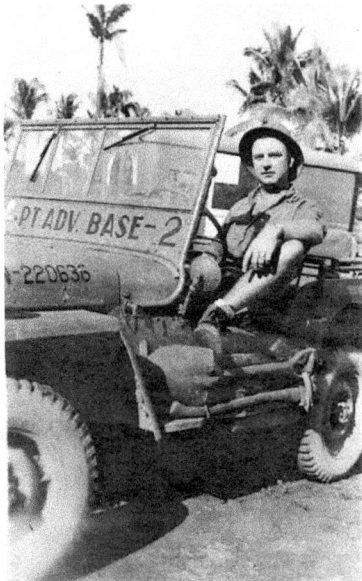

This jeep is marked as Advanced PT Base 2. Notice the driver is wearing what was called a pith helmet which was actually an American fiber helmet, made from pressed fiber. The helmet did a wonderful job in keeping the hot Pacific sun out of one's eyes and off the wearers head. (Frank J. Andruss Sr.)

Base Force Torpedomen roll the MK-XIII after bodies from the torpedo dump on the island. The business end or forward body of the torpedo had a 600-pound high explosive war head and was usually kept in a different area of the torpedo dump. The torpedo when fired could travel to the target at roughly 33.5 knots. The MK-XIII weighed in at 2, 216 pounds and would replace the heavy steel MK-18 torpedo tubes first used on the boats. (Frank J. Andruss Sr.)

Unknown PT Boat and her crew have managed to trade beads for Tobacco products to the locals. Here they are showing off the handmade beads, which were usually sent home to a loved one, or came home after the war. These barter sessions continued throughout the Pacific with some very nice handmade items. (Associated Press)

This was the fuel and dispensing area on the base. The area was built by the 40th Construction Battalion. As far as the eye can see, there are stacked fifty-five-gallon steel drums. A truck has pulled into the area and is fueling up as gasoline is pumped from a large dispenser using a pump system to dispense the fuel without the need to hand crank. (US Navy Seabee Museum)

Seabees from the 7th Naval Construction Battalion are working on the Seabee theater at the base. When finished this would be used by all Navy personnel including those from the PT base on the island. (US Navy Seabee Museum)

A makeshift shower system and shaving stand built by the Seabees for the PT boat base on the island. Notice the fifty-five-gallon drums that carry water that would cascade down to the waiting recipient for washing away the sweat and dirt of the Pacific. (US Navy Seabee Museum)

16. A typical scene that was repeated by the US Navy CB units, all over the South Pacific. Permanent docks are built from the shore that can handle cranes and other materials that are connected to floated docks for the boats to tie off. This set up makes it easier for materials to be brought to the boat and for Torpedoes and Engines to be added or removed. (US Navy Seabee Museum)

TULAGI

It was in October of 1942 that a detachment of fifty-nine men of the 6th CB Battalion was sent from Guadalcanal to Tulagi to build a PT boat base at Sesapi. Later this would increase to 133 officers and men. To begin work they constructed an emergency outlet channel for Tulagi Harbor, by dredging and blasting. By doing this it would avoid PT boats becoming bottled up by enemy warships. Two PT Boat floating dry docks were assembled from pontoons and a 500-man camp was set up. Another item that was urgently needed by the PT base was a radio shack. The ships moored along the bank of the Maliali River made it impossible to get a good signal as the high hills blocked the radio transmission between the PT tender Jamestown and boats on patrol. With help from the Marine headquarters on Tulagi, they would relay messages from coast watchers and air patrols but could not maintain constant communications with the boats at sea. With the arrival of more equipment, the PT Boat men were able to set up their own radio shack at Sesapi, using a Japanese generator.

They would also provide several carpenter details to help with the maintenance and repair of the boats. By August of 1943, the facilities at the Sesapi base would expand to include shop facilities, and storage areas that would permit major PT Boat overhaul. A repair and service unit was set up that was able to support 40 PT boats in combat operations. The Seabees would also construct three small wharves for the boats. Much of this later work would be done by the 27th CB Battalion.

In addition, PT Boat facilities would be constructed on the island of Macambo, with base housing at Calvertville on Florida Island. They had an existing concrete wharf on Macambo that needed repair but was still serviceable. They had to build torpedo overhaul and storage facilities. During December 1942, PT boats from Motor Torpedo Boat Squadron Two were based here. On December 31, four PT boats of Motor Torpedo Squadron Six arrived and immediately went into action against the

Japanese Tokyo Express. In March 1943, the third squadron of RON 6 arrived. In July of 1943 PT Boat Squadrons 1, 3, and eight were using Sesapi and the Macambo bases. The Seabees would also construct two 1000-barrel tanks for aviation gasoline at Sesapi and eight 1000-barrel tanks at Macambo that had loading line to the dock.

The base at first was a simple combination of tent structures that could quickly be set up to house sleeping quarters and base force shops. The base just off the south coast of Florida Island was within sight of Guadalcanal. The base at Sesapi did offer some good things as it was isolated from the main harbor, so the boats did not interfere with other naval activities. It also provided pretty good protection from wind and heavy seas. (Frank J. Andruss Sr.)

At the base heavy rains would turn these areas into mud holes. Infested trenches with mosquitoes would made everyday living a nightmare. Boat crews would sometimes sleep in tents but would often choose to sleep on deck. The rain and humidity made for uncomfortable sleeping with relief coming when the boats would be out in the water. (National Archives)

Seen here is this captured Japanese 13mm dual purpose anti-aircraft weapon that was utilized and set up to help defend the island against Japanese plane attacks. (National Archives)

This camouflaged enemy truck seen amid the ruins, caused by Navy gunfire on Tulagi was captured by Marines. Later it would continue to operate transferring supplies on the base, being used by the PT Boat base force. (National Archives)

Wooden A frame built by the Navy Seabees. Right at the water's edge we can see several types of duty boats pulled into shore. This framed building was stocked with many different supplies needed for the base. (National Archives)

A look at one of the water purifying stations at the base. Seabees would construct these drinking locations on the base so that personnel would be drinking safe water. All bases required enough drinking water not only for drinking but basic hygiene, laundry, cooking and a host of other requirements. Providing a source of potable water at advanced bases remained a major issue throughout WWII especially on coral atolls and major operating bases throughout the Pacific. (National Archives)

A look at the PT Boat Nest at the docks. This was taken at what would be known as Sesapi in Tulagi Harbor. In the photo we see a combination of Elco seventy-seven-foot boats and the newer Elco eighty-footers. Also notice to the left the very important dry docks used to remove the boats from the water to perform maintenance on the hulls. (Frank J. Andruss Sr.)

A native grass and bamboo structure made up the First Aid station on Tulagi. This First Aid area would tend to all casualties. Medicine, food, blankets and emergency care were all handled here. Severe cases would be shipped out to better equipped medical facilities. (PT Boats Inc.)

Wonderful crew photo of *PT-61*. Ensign Ken Prescott, the boats Executive officer is on the far left with the.45 Pistol on his hip. Ensign Joe Kernall, the boat's Skipper is next to him in cockpit with his back to the wheel. The boat was active in the Solomon's campaign, engaging in many strenuous night actions with the Tokyo Express in the defense of Guadalcanal (Ens. Kenneth Prescott)

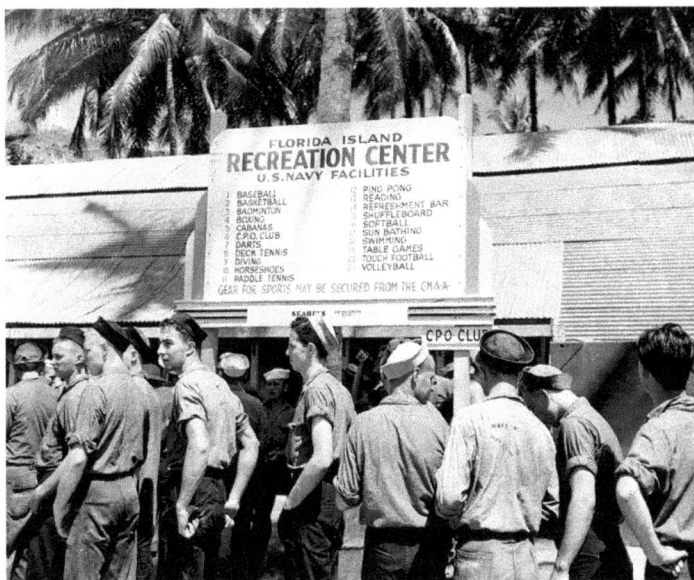

Across from the Harbor from Sesapi was the PT base known as Calvertville located on Florida Island. Facilities were constructed here that included new barracks for the boat crews. This base was named for Commander Allen P. Calvert, commander Motor Torpedo Boat Flotilla one. The Florida Island base was set up as Sesapi had started to outgrow itself. Engine shops, torpedo shops, and warehouses were enlarged, and a recreation center and officers club were built. In addition, PT facilities were constructed on the nearby island of Macambo. (National Archives)

Lt. Charles Tilden checking out maps and his charts, below deck of *PT-43*. On January 11, 1943, Japanese destroyers under the cover of a rain squall were missed by two of the PT scout groups, but later were spotted by scout group one. Three Japanese destroyers were spotted off the coast of Guadalcanal, and the three PT boats 112, 43, and 40 attacked. Sadly *PT-112* was sunk and *PT-43* so badly damaged it had to be abandoned. *PT-43* was later sighted beached on the Japanese held portion of Guadalcanal and was destroyed by gunfire from a New Zealand corvette to prevent capture. (National Archives)

Unknown Elco seventy-seven-footer heads into the Sesapi PT boat base in Tulagi Harbor. These little boats would see the majority of the action in the early part of the war in the Solomons, gradually being phased out by the larger Elco eighty-footer. (Frank J. Andruss Sr.)

Supplies were the lifeline to the men that served on these bases. Here transports have pulled into the dock area at Tulagi. (National Archives)

Two crew members take advantage of the sun as they enjoy the ride just off the Tulagi coastline. This is one of the Elco seventy-seven footers that was stationed at Sesapi and most likely was on the way home. (Time Life photo)

Coast watchers provided valuable intelligence on Japanese movements and troop dispositions during the fighting in the South Pacific. Control of the Solomon Islands and New Guinea was vital to Allied plans for a counteroffensive in the region and to safeguard Australia against invasion. Here Martin Clemens and his scouts provided the US Marines much assistance with continuous raids on Japanese supplies and radio reports of the enemy's position. (National Archives)

Elco seventy-seven-foot PT boats skirt the coast of Tulagi. Early in the war it was these boats that did battle with the Japanese in an area known as the slot. Leaving their base just before nightfall, the boats would wait in ambush for the Japanese destroyers to head past, then sneak out to try and get close enough to launch torpedoes. They operated in pairs, maintaining as stealthy posture as possible until they could attack. (Time Life photo)

This lone Elco eighty-foot PT Boat hugs the Tulagi shoreline in 1943. The boat is tied under overhanging trees, which make it a very difficult target to spot from the air. (National Archives)

True to tropical traditions, native and PT crew trade food for tobacco and beads at Tulagi base. At the dock is one of the Elco seventy-seven-foot PT boats. (National Archives)

A wonderful, majestic look at Tulagi looking East from hill 281. Taken in 1944, one can see the buildup that has taken place since 1942. When seeing photos such as this, it is hard to believe that war came here and so much fighting took place. (National Archives)

This ship's cook is dressed in his summer white uniform, obviously for an upcoming inspection. Perched on the top of his white cap, is his friend, a local parrot that he has trained while at the base. It was not uncommon for PT Boat sailors to have mascots such as birds, monkeys, or small local dogs. (Time Life Photo)

Sometimes getting to the underside of the PT boats was tricky business without the aid of a dry dock. Here an underwater diver is about to check out one of the boats checking the struts, shafts, and wheels. Getting a good visual of the problem made things easier for repairs. (Time Life photo)

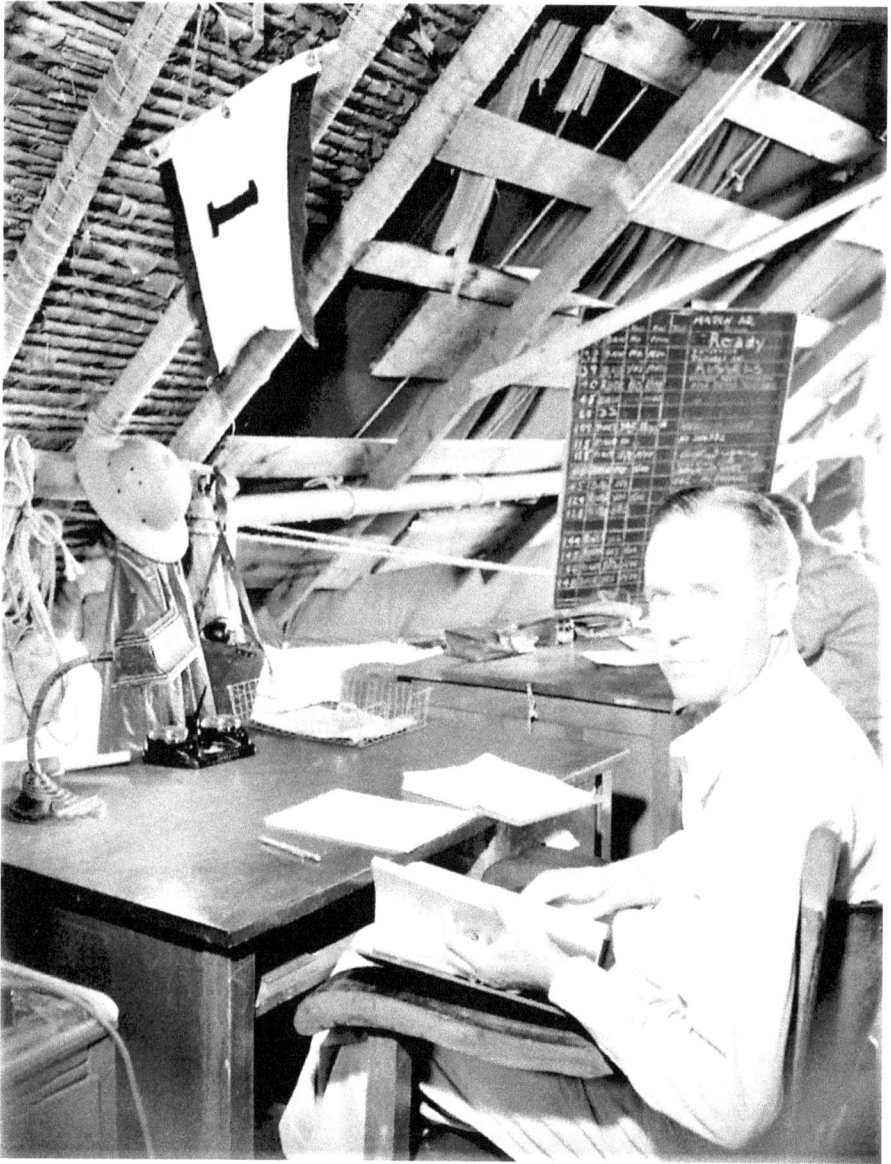

Cmdr. Allen P. Calvert USN was the Commander Motor Torpedo Boat Flotilla one. From December 1942 until November 1943, he was in command of Motor Torpedo Boats, Solomons Area, South Pacific. For meritorious service in that assignment during the period December 15, 1942, until February 1943, he was awarded the Distinguished Service Medal by the War Department. After the war he would go on to command the Pacific war fleet in San Francisco. Calvert would have a long career in the Navy becoming Rear Admiral. (Time Life photo)

Under the palms of beautiful Florida Island, these sailors become carefree youths again as they relax at the recreation center. Here they had facilities for sports, beer, and soft drinks. One of several centers that were sponsored by the US Navy to offset the strain of combat duties, the installation was built and run by the Seabees. (National Archives)

RUSSELL ISLANDS

On February 21, 1943, a combined landing force, Code-name: Operation Clean-slate was the Russell Islands Occupation. In this unopposed landing, useful as a dress rehearsal for the invasion of the Central Solomon Islands, Captain Ingolf N. Kiland, USN, commanding Task Unit 76.7.2, landed the US Marines Third Raider Battalion and 10th Defensive Battalion detachment along with the US Army 43rd Division on Russell Islands. The islands became an important staging point for the invasion of New Georgia, Northern Solomon Islands, Bismarck Islands, and even for the invasion of Okinawa in April 1945. The Russell Islands consisted of several islets and two main islands, Mbanika and Pavuvu.

By late January 1943, Lt. Jack Searles was returning from a Rest and Relaxation (R&R) trip with Squadron 3(2)'s original officers and he was requested to meet with ADM Bull Halsey at his headquarters in Noumea, New Caledonia. At this meeting the Admiral informed Searles that he and his PT boats would be moving north to the Russell Islands to set up a new PT boat base that would be needed to support a future invasion up north. The Russell Islands were not far, only fifty miles northwest of Guadalcanal. On 1 February 1943, Lt. Jack Searles, relieved Lt. Hugh M. Robinson as Commanding Officer of Motor Torpedo Boat Squadron 3(2), at Tulagi.

In March 1943, Lt. (jg) Bryant Larson, skipper of *PT-109*, was the first to go up to the Russell's on a night mission to rendezvous with a Marine Scouting party and pick up their CO so he could report to personally to his chain of command, on Guadalcanal. Soon afterward Lt. Jack Searles took several boats from RON 3 (2) up and entered Renard Channel and found a suitable location to set up the new PT base, on Mbanika Island.

The officers billeting and headquarters was in The Lever Brothers Plantation house. Before the war started, the island was the location of a large pineapple plantation.

The PT boats started patrolling from this location immediately, but except for the few Japanese air attack, there was no real action. The PT boat squadrons used this time to train new personnel that arrived to relieve the original Tulagi crews. Also, new squadrons arrived from the states, the Russells were an excellent training area for them too, before, moving up the Solomon Islands chain. Later, once the base was set up and the PT squadrons had set up shop, Adm. Halsey, stopped by on a quick tour. After meeting the men, Halsey suggested the new base be named "Searlesville."

The base sort of became a popular place, one day Ens. Ken Prescott, had the duty and was the Officer of the Day (OOD), he noticed a seaplane had landed in the lagoon, and motored toward the dock. He had received no word that a seaplane was going to land, so this was unauthorized. Ens. Prescot grabbed two armed guards and quickly made his way to the dock to find out what this was all about. As he arrived a sailor was opening the side access hatch and out jumped a dark-haired man in his early 50's, wearing khaki's with no rank insignia, to Ken's surprise, the man was Undersecretary of the Navy James V. Forrestal, he spent a few days with the PT Men, before moving on his "fact finding" tour, Later the United Service Organizations (USO) troupes came through to entertain the servicemen.

As previously mentioned, as time passed, the Russell's became a major staging and supply area for future invasions, in the island-hopping campaign. The Seabees arrived in the Russell's and started building the facilities that would support these operations. One group that contributed to the expansion of facilities was the 93rd CB Battalion. Once they erected their camp, they built an Advanced Base Construction Depot (ABCD), which included an equipment depot and sawmill, when these were in place, a few projects were completed by the 93rd. Roads were built, other camp facilities were enhanced, warehouses and storage areas were constructed, and existing dock facilities were improved.

Then followed the first major overseas assignment, construction of naval hospital, naval hospital dispensary, naval Recovery hospital, naval Mobile Hospital-MOB 10. M'Banika Island was to become a major supply and causality recuperation base for Armed Forces personnel wounded and injured during the island-hopping campaign in places like Rendova,

Bougainville, Kwajalein, Tarawa, Eniwetok, Saipan, Guam, Tinian, Peleliu, Iwo Jima, and Okinawa.

The airfields were constructed and used to attack Japanese forces on New Georgia and Munda. Banika Field: was used by Navy, Army Air Forces (USAAF) and Marine Corps squadrons.

VMF -121, VMF 214, and VMF(N)-531 were three Marine fighter squadrons that operated from this field. Renard Field: was used by USAAF Bomber squadrons and Navy VB-140 and VB-148 (both equipped with the Lockheed Ventura PV-1). Renard Sound Seaplane Base, which was located near M'Banika Field, in the channel to the east of the runway, Navy PBY Catalina (Black Cats) operated from this base, and often coordinated with the PT boats in anti-shipping patrols.

The Lever Brothers Plantation House Officers Quarters Searlesville, Mbanika Island, Russell Islands 1943. (Capt. Kenneth W. Prescott, USNR).

Volleyball game Searlesville 1943. The PT men had plenty of sports and recreation at this base, Baseball was another favorite. Only once did they play football, full contact, and that was later forbidden after one of the players had his leg broken.

USS Hilo (AGP-2) with PT boats from RON 3(2) and RON 10 Sunlight Channel Searlesville 1943.

PT-116 RON 6 aboard a transport ship. Eventually, this boat would be operating out of Searlesville.

PT-61 RON 3(2) Chief Burney Fisk teaches the crew how to send messages to the other boats using semaphore signal flags, Sunlight Channel Russell Islands May 1943. (Capt. Kenneth W. Prescott, USNR).

PT-61, *PT-48* and *PT-109* moored in the bushes, Russells May 1943. (Capt. Kenneth W. Prescott, USNR).

PT-61 crew color photo, Searlesville May 1943. (Capt. Kenneth W. Prescott, USNR).

RENDOVA

The Rendova PT Boat Base was very well thought out and was an outstanding example of careful selection of site, proper planning, and efficient operation. Rendova PT Boat Base (Lumbaria Island odd   City) was located on Lumbaria Island inside Rendova Harbor bordering Lumbaria Island off Rendova Island. This base was established after US forces had landed at Rendova on June 30, 1943. Starting in early July 1943, Rendova PT Boat Base was used by the USN as their primary and most forward PT boat base in the central Solomon Islands to support the New Georgia campaign.

The base was under the command of USN Commander T. G. Warfield and was also known as Lumbaria Island PT Boat Base. At this location was based Motor Torpedo Boat Squadron 9 and Motor Torpedo Boat Squadron 10.

In early July 1943 this base was renamed "Todd City" in honor of Gunners Mate 3/c Merwin Kenneth Todd *PT-162* RON 9, the first PT man killed at Rendova, who was buried at sea. By 1944, the Seabees had constructed three floating dry docks, two floating piers, wood working, engine, torpedo, radio and electrical shops. They also had a 40 x 100 administrative building and undercover storage for spare parts. They could perform complete boat and machinery overhaul work here, although they would send major engine overhaul to Espiritu Santos. Boat hulls could be scrapped at the base and left on dock to dry out before re-painting them. After engine overhaul the boats would be tested for efficiency. A complete schedule of work to be accomplished on each boat was laid out with division of work between base and personal, the order of the day. The base had a reasonable supply of Elco boat parts on hand but at times suffered shortages of props, shafts, mufflers, batteries, exhaust stacks and cutlass bearings.

A great look from the air of PT Boat Base 11 located at Rendova, on Bau Island. This was the Navy's primary and most forward PT boat base in the central Solomon Islands to support the New Georgia campaign. Notice in the photo they have dry dock areas and inland several large warehouse buildings. Also they have a very good dock system, capable of holding several PT boats as needed. (National Archives)

A look at Squadron 9 anchored off Lumberi Island, the site of Todd City PT Boat Base, which was the first PT base at Rendova. This base served as an operational center for daily patrols and the headquarters of PT boats, Rendova. One of the squadron's members (GM2/c Merwin Kenneth Todd) was the first PT boater to die there—and it was he who lent his name to the base. (National Archives)

Local Natives help with unloading this US landing craft. It was not uncommon that the Natives would be enlisted to help with many of the local chores at Bases in the South Pacific. This labor helped to free up Navy enlisted personnel who could be used elsewhere for other important jobs. (National Archives)

PT-180 and *PT-183* anchored side by side at Rendova. These boats would serve with RON 11. *PT-183* was damaged on August 21, 1943, by a strafing Japanese float plane while patrolling near Turovilu Islands in the Solomons. Both boats would be transferred to the Southwest Pacific in June 1944. (QM1/c John C. McHenry)

Overhead look at the other PT boat base. Known as Base 11, it was located in a cove on Bau Island inside Rendova Harbor off Rendova Island in the Western province in the Solomon Islands. To the west is Pau Island and Lumberi Island, the PT Base known as Todd City. (National Archives)

Lt. Cmdr. Jack E. Gibson checks out the remains of this downed Japanese plane at Rendova. He served with PT Squadrons 4, 10, and 11 during the war. His first combat with PT boats was with RON 11 while based at Rendova. He served on the base at Vella La Vella as Commander of Squadron 10 and would set up the PT boat base on Green Island. (PT Boats Inc.)

This is the Squadron 9 honor board at Lumberi, showing the names of those PT boaters killed in the line of duty. To the left is Pharmacist Mate 1/c William J. Lawrence and on the right Doc Horton (National Archives)

The pier at Todd City was named after Harold W. Marney Momm 2/c who was killed on *PT-109* in August 1943. Marney grew up in East Springfield, Massachusetts attending Trade High School, where he studied auto mechanics. He joined the Navy at age 17, a month before the attack on Pearl Harbor. (National Archives)

Two boat crew sailors at the base have the tedious task of cleaning ammunition for the boat's weapons. While on patrol, ammunition was exposed to saltwater spray, and upon return would have to be discarded and replaced, or in this case cleaned and ready to go for the next patrol. (Frank J. Andruss Sr.)

Sign at Todd City in honor of Lt. Joseph D. McLaughlin. He was the second officer on *PT-154* under skipper, Lt. (jg) Hamlin D. Smith. *PT-154* was conducting an anti-barge patrol with *PT-155* when a three-inch shore battery opened up. One of the rounds hit the after-body of the port forward torpedo where it exploded killing McLaughlin and QM2/c Arthur Schwerdt. The boat captain Hamlin and six other crew members were injured. (National Archives)

Boats from Squadron 11 tied up at Rendova. Notice the tarps that are set up on the boats bow. This was so that work being conducted would help shield sailors from the tropical sun, which could be brutal in the heat of the day. The hot sun would cause metal parts to become too hot to handle. (National Archives)

Crew members from *PT-157*, Squadron 9 are heading home at Rendova while operating in support of the New Georgia operation. Notice the "Aces & Eights" insignia is painted below her number on the front of the cabin. Her starboard side mount now has a 20mm Oerlikon cannon in place of the usual twin.50 caliber machine guns. Her skipper Lt. William F. Liebenow is standing in the center with white tee shirt on. (Lt. William F. Liebenow)

Crew from *PT-122* takes time to pose for the camera. This boat would serve with Squadron 6 arriving in the South Pacific in time for its boats to participate in some of the last actions with the Tokyo Express at Guadalcanal. While serving with Squadron 8, she was credited with the sinking of Japanese Submarine I-22 on November 12, 1942, off the mouth of the Kumusi River, fifteen miles southwest of Buna, Papua, New Guinea. (Frank J. Andruss Sr.)

Crew members from *PT-176* take a bit of time to play in the mud. Heavy rains caused flooding at the base and mud was the order of the day, after a steady diet of rain. Roads would be tough to drive on in some instances and it would take some time before the mud would dry out. (Bob Hart MoMM1/c)

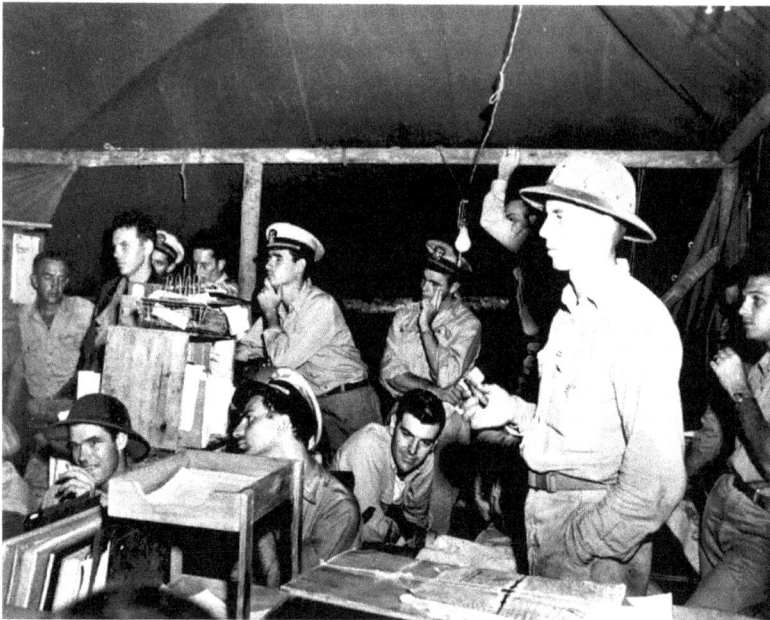

The look of concern as the boat captains have gathered in the operations tent for briefing. Here they will learn which patrol areas they will cover, and where Native coast watchers have spotted Japanese shipping movements. These are tense times as the little boats and their crews know that they could run into stiff resistance from the enemy. (National Archives)

Some of the locals occupy the base work truck. They worked side by with the base force, performing manual labor and whatever was needed to insure smooth operations. This type of labor was all-important as it helped free up men that were being used elsewhere. (GM2/c Robert Douglas)

LEVER HARBOR

Following Rendova, the PT boats moved next to Lever Harbor on 'The Slot' side of New Georgia Island to begin the even more difficult task of attacking the well protected Japanese barges. One good thing about this area was that it had existing buildings that were built before the war, by the Lever Brothers which was a soap company. These buildings were constructed for their staff and productivity facility. PT boats when they moved in docked their boats under over hanging trees just a few feet from shore. The crews would build ramps from coconut logs so they could step off their boats and onto land. The crews were fed at least one meal per day in the Lever Brothers production facility, which was turned into a chow hall. At this base PT crews could get their clothes washed by friendly natives, who traded this chore for Navy fish rations.

The Lever Harbor boats, which had their first barge action on August 3, 1943, engaged forty-three barges from then until the end of the month, of which two were sunk, one was forced to be beached, and 8 to 16 were hit with possible damage. PT boats also continuing to harass enemy logistics flow, quite successfully, it seems. Captured Japanese reports refer to the challenges presented to barge operations by the PT boats.

Seabees working at Lever Harbor have established buildings with tent structures and continue to work making better living conditions. Camouflaged netting has been strung up between palm trees to help conceal the area from Japanese planes. Two Seabee's are digging what appears to be a trench. (National Archives)

Existing buildings such as this one, were built before the war, by Lever Brothers a soap company. The buildings were constructed for their staff and productivity facilities. It made living conditions a bit better for those at the base. (Frank J. Andruss Sr.)

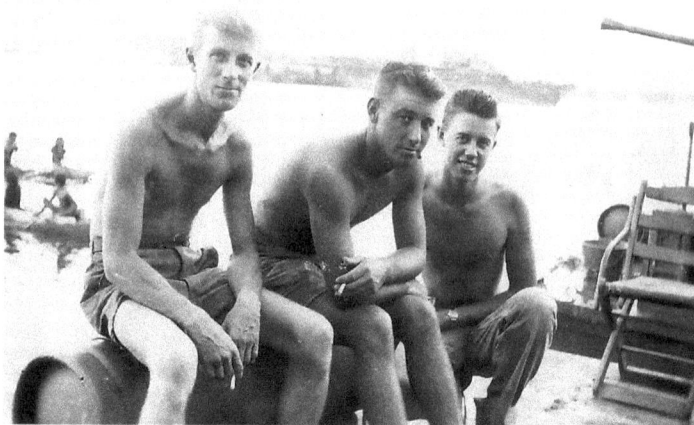

Unknown PT boat crew takes time for a smoke on the boat. Typical dress of no shirts and shorts can be seen in this photo. To the left, natives in their outriggers are approaching the boat. As with most crew, they are skinny and one day hope to get back to Mom's cooking. (Frank J. Andruss Sr.)

One of the main adversaries of the PT boats in the South Pacific were the Japanese barges and landing craft. Here this landing craft has been hit and beached. Tactics for the PT boats was to make runs at the barges bringing all weapons to bear, hitting them quick and hard. Sometimes it was not that easy as the Japanese had larger caliber weapons onboard and sometimes lined the boats with metal and sandbags. It was for this reason that the PT boats began to use hard hitting cannons. (National Archives)

Typical Native living area on Lever Harbor. It is most likely that more than one family lives here. If not for war, this makes for a beautiful view out to the Harbor. Surprisingly the thatched roofs really did well in keeping the elements out. The tall roof provides a space for heat to escape, allowing cooler air to be drawn in from the windows in the walls below. (National Archives)

PT-154 heading into Lever Harbor. She would serve with Squadron 9 under Lt. Cmdr. Robert B. Kelly. Notice she is carrying a 20mm Cannon in her starboard side turret which replaced the twin.50 caliber machine guns. (Frank J. Andruss Sr.)

This PT boat is having one of her 4M-2500 Packard marine engines removed. Navy specs called for these gasoline engines to be replaced after 600 hours, but many times under war conditions, this did not happen. Shortages in supplies for the engines sometimes called for the Motor Machinists mates to baby the engines. It was not uncommon for some parts being taken from other engines and placed into the boats that would be heading out for patrol. (Frank J. Andruss Sr.)

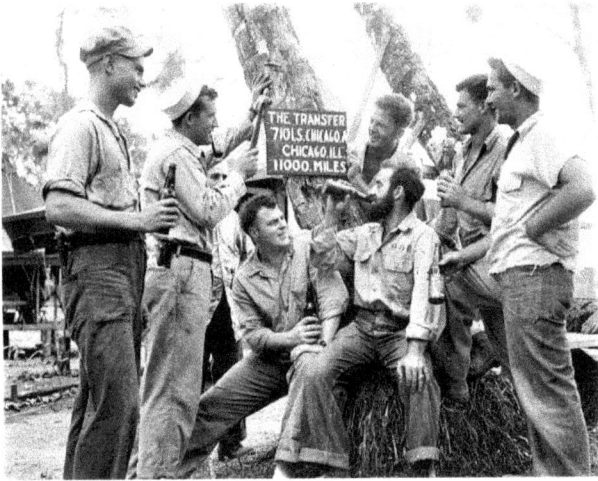

A scene duplicated at many of the Pacific Islands, these sailors have gathered for some good conversations and more importantly their ration of beer. As the sign mentions they are a long way away from Chicago. (National Archives)

PT-187 about to receive a newly conditioned 4M-2500 Packard marine engine. Each PT boat had three of these super-charged gasoline engines that produced some 4,500 Horsepower. They were a lightweight engine that had their own built-in reverse gear produced by Joe's Reverse Gears in Connecticut. These engines had a cruising power rating of 900 HP at 2000 rpm. and what was known as an emergency power rating of 1,350 HP at 2500 rpm, enough to drive the boat at some forty-one knots. (Gordon Goosela MoMM2/c)

A look at this boat's Motor Machinist Mates performing daily maintenance on the boat's engines. Hot, dirty work made things very difficult but for them, but they always made sure that boats heading out for patrol were ready for action. (Frank J. Andruss Sr.)

VELLA LA VELLA

The occupation of Vella La Vella would mark the close of the central Solomons campaign. Seabees would establish a small airbase and naval base here. The 58th CB battalion landed here on August 15, 1943, in which under heavy bombing would begin to construct nine miles of road and erect tents for quarters. The next construction would be a dispensary and sick bay, which consisted of four underground shelters, each with a capacity of four beds, and an underground operating room. During the next several months, the channel through the reef was deepened to allow PT boats into the lagoon. The jetty was improved, and a camp was set up at the navy base with a marine railway for the boats, and construction of a boat repair locker. Early on the base consisted of one rusty sheet iron building and three native grass huts. The building was used for a few spares for the boats, torpedoes and guns. The three huts were used for storage, with one being the armory with all the ammunition. The dock was very small and only one boat could tie up to it. This is why a strict rule of no more than two boats would be tied together at the dock at any one time.

The boats fueled from 55-gallon drums using a gasoline pump set on a drum until it was pumped empty. Water in the fuel was always a problem, and the gasoline was filtered through a funnel with a chamois sheep skin. The fuel drum storage of about 600 drums was in a swamp next to dock area. Fuel was brought in about once a week by Landing Craft, Tank (LCT). The ramp would be lowered in about waist deep water.

All hands including the boat crews not busy would roll the drums out into the water and float them into the swampy storage area and stand them up together. They would load the empties back on the LCT. The base personal tents, communications tent, officers' tents, sick bay, and galley were set up about 200 yards along a coral path back in the woods. PT crews ate outside the galley in the open under coconut trees on picnic style tables, rain or shine. Food was pretty bad, although it was not the cook's fault, they just didn't have the correct supplies on hand. On Sundays a chaplain from the air base would come up to the PT base to conduct services. PT boat operations at this base would end the last of the construction battalion on Vella La Vella would leave this base in January of 1944. PT boat operations would cease at this base as they began moving up to Treasury Island.

LST-485 was laid down on December 17, 1942, at Richmond, California, by Kaiser Inc. She was launched on January 9, 1943, and commissioned on May 19, 1943. Assigned to the Pacific area, she is seen here at Vella La Vella sometime in September 1943 as she begins dropping off supplies. (National Archives)

PT-106 heading back in from patrol. Patrols were long and tedious as the boats operated in the black of the night. Many nights were spent running at slow speed, where eyes could play tricks on you, one minute seeing something and the next nothing. Tired bodies from endless patrols would often sack out before pulling into base. (Frank J. Andruss Sr.)

Part of the crew of this unknown PT boat take time to relax on the bow. Rest was very much welcomed after long patrol's and working to get the boat in shape. These sailors are doing what they can to relax during the day, which could cause a stupor given the heat and humidity which was a huge part of the Pacific. (PT Boats Inc.)

Standing out in front of the many tent structures that would make up the PT base at Vella La Vella. As time moved on, structures and other buildings would make things a bit more bearable, although not much as boat crews ate outdoors under coconut trees in picnic style tables, rain or shine. (PT boats inc.)

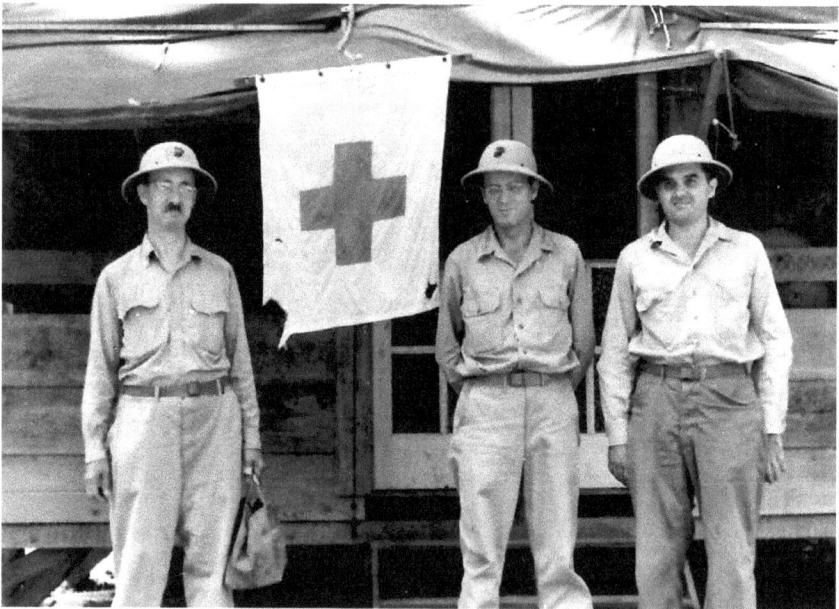

Left: Cmdr. Holman, Lt. Cmdr. Tobin and Medic Hauck made up the sick bay team on Vella La Vella. Injuries and or sickness were treated here, with more severe cases transferred to the rear. (Frank J. Andruss Sr.)

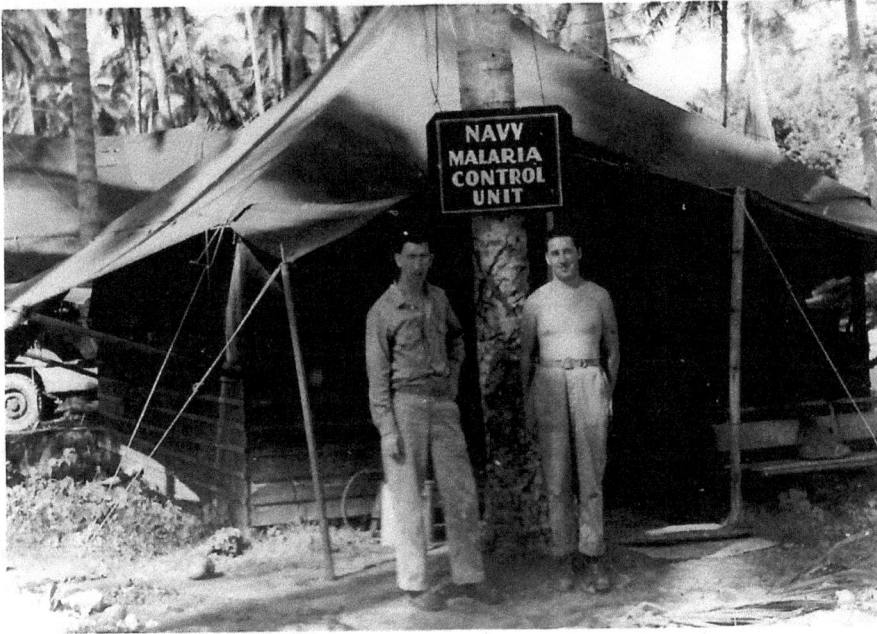

Although the sign might appear humorous to the reader, this disease, while not typically fatal to the infected sailor, would take him out of action for a prolonged period just as surely as if he was wounded in battle. The disease itself is caused by a mosquito-borne parasitic protozoan that attacks the red blood cells and liver of the infected person. The damp, swampy environments encountered on many Pacific islands were ideal breeding grounds for the anopheline mosquitoes that transmit the disease. The most used drug to combat malaria was quinacrine, better known by the brand name Atabrine. While effective, Atabrine had several side effects, including a tendency to turn a soldier's skin bright yellow. (Frank J. Andruss Sr.)

A rare look at *PT-60* (in her gunboat configuration) as she moves out slowly from Lambu Lambu Cove on Vella La Vella. Notice there was very little room to move in and out of this cove. To the right you will notice the gasoline dock which was destroyed in a gasoline fire on December 14, 1943. At the helm of the boat is Ensign Leonard Thom who was the executive officer on *PT-109* at the time she was lost. (Frank J. Andruss Sr.)

This sparse shack set up made for the medical supply office. The tent structure made up the roof and was drawn down to keep the rain out of the building. Sometimes because of high humidity and exceedingly heavy rainfall, supplies at many bases were lost or deteriorated beyond use as a result of a lack of proper storage facilities. Labels became unreadable or fell off the containers entirely, leaving no choice but to destroy the contents. Metal components of equipment became rusted or corroded, and fungus grew on certain items. In this photo it is hard to tell if screens have been added to keep out the bugs. (PT Boats Inc.)

A common scene that was repeated throughout the war in the Pacific. Natives worked with the allies in a variety of ways. The most important was their ability to watch Japanese troop movements. Natives worked with PT Boater's loading and unloading gasoline drums and doing various tasks at hand. (National Archives)

PT-116 heading out from the base at Vella La Vella. She would serve with Squadron 6 and Squadron 10. It appears that her mast has either been removed or is in the stored position. Many times masts would be stored so that the boats could get closer to the shore and overhanging trees. (Frank J. Andruss Sr.)

Tribal Natives from Vella La Vella stand in their dress, while two officers look on. One Native is carrying a hatchet while the other looks to be carrying a spear. (National Archives)

Mess area conditions on the base were nothing more than tents strung up between the many trees on the island. This area as well as others was set up roughly 200 yards from shore. (PT Boats Inc.)

Some of the crew of *PT-108* showing their captured Japanese Flag. Known as a good luck flag, they were traditionally a gift for Japanese servicemen deployed during military campaigns of the Empire of Japan. The flags were usually signed by family and friends often with short messages, wishing the soldier victory, safety and good luck. (Frank J. Andruss Sr.)

At Vella La Vella as were many of the Pacific based Islands, the shower area for the men was outdoors. Seabees rigged up a simple wooden platform with overhead shower rings connected to stored water drums. Simple but effective for cleaning up, after working in the heat and humidity of the day. (National Archives)

Treasury Island

The Treasury Island PT boat base was some twenty-eight miles south of Bougainville, and roughly twice that distance northwest from the former Vella La Vella base. This site was chosen as an advance air and naval base to neutralize the Japanese strongholds on New Britain, New Ireland, and Bougainville. These islands consisted of Treasury or Mono Island and Sterling Island. Blanche Harbor, between the islands had a deep channel, half a mile wide through its eastern entrance. Stirling PT Boat Base (Treasury Base) was located on the shore of Stirling Island with anchorage in Blanche Harbor.

On October 27, 1943, Company A and twenty-five men of the headquarters company of the 87 CB Battalion landed with the 8th New Zealand Brigade, in the first wave making the assault on the island. These operations were carried out under enemy high-level bombing. Mortar, and machine gun fire.

After enemy resistance was eliminated, the Seabees went to work improving beaches for landing craft, built roads, gun positions and built a wharf for the PT boats on Sterling Island. Later the Seabees would build a fuel station for the boats with three pontoon dry docks. A crash boat pier and a small-boat pier were also constructed. PT boats based in the Treasury Islands group helped to protect Allied forces landing at Torokina, while the radar site at Soanotalu played an important part in the success of that operation. PT boats from this base patrolled against Shortland and Bougainville during December 1943 through May 1944.

Later in 1944 after combat operations slackened, this base was used for "combat training" by three PT boat squadrons, making nightly patrols on the bypassed area on Bougainville and Choiseul, but rarely encountered enemy barges.

Treasury was a beautiful island. All along the shore where the boats berthed there were huge trees with horizontal limbs that hung out over the water. You had to walk one of those limbs to get ashore. The water was deep enough along shore that you could tie the boat under these overhanging limbs. The water was clear, and you could see 20-30 feet deep. Some of the PT crews would clear a place on shore at their berth for a place to cook and eat. C.J. Willis on *PT-242* and the crew had set up two tents on shore where they had a propane gas field range stove where their cook, Bill Knapp, cooked most of their meals. This was the only base that they were at in the Solomon's where they cooked and ate at the boat. The other bases they were at, they ate at the base chow hall.

Boat crews slept in the tents on the shore, while others would sleep under a tarp on the bow of the boat. One night the guys on shore all came charging on the boat. There was even a small earthquake that shook boat crews awake and cracked some tree limbs around their tents. PT crews would cut the heads out of steel drums and place them around the tents to catch rainwater for bathing. By late 1944 the base was far behind the active combat zone, and it was disestablished in March 1945.

PT Boat sailors raved about Treasury being a beautiful place. The water was deep enough and so clear that one could see 20 to 30 feet deep. All along the shore where the boats were berthed, there were huge trees with horizontal limbs that hung out over the water. The water was deep enough along shore that you could tie the boat under these overhanging limbs. (PT Boats Inc.)

This photo shows the communications shack at the base. Radio equipment was kept here that could transmit and receive important messages. Information gathered over the patrol areas and beyond were most vital so that Commanders could quickly estimate the situation and make any tactical decisions. Coast watchers would relay important messages which would point out any Japanese movements on water or land. (PT Boats Inc.)

A majestic photo as this unknown PT Boat heads into the Treasury base after a long night's patrol. Their work would not be done however, as the boats needed to be fueled, ammunition replaced, and any other problem that arose while on patrol, taken care of before the crew could knock off for some much-needed rest. (Frank J. Andruss Sr.)

A typical scene that was repeated all over the South Pacific PT bases. These natives from the area have paddled over to the boats to trade, and possibly share information on Japanese movements in the area. At the outset of war, as Japanese troops invaded the Solomon's group, Native Coast watchers went into hiding in the bush and began reporting on enemy movements to Allied headquarters. The Coast watchers' work was so significant in winning the Solomons Campaign that US Admiral William "Bull" Halsey, commander of the South Pacific Area, proclaimed that the Coast watchers saved Guadalcanal. (National Archives)

Using a jeep to get around the base was important, providing that base had roads which could be utilized. Jeeps did everything around the Islands. They performed cargo hauling, communications work including laying telegraph cable, and recovery work, and were frequently used as transport vehicles for executive staff. They worked as front-line ambulances if needed. (Frank J. Andruss Sr.)

Wonderful overhead look at the motor torpedo boat base at Treasury Island. From the air the base looks very secluded. Notice to the left, you can see one of the PT boats in dry dock. (National Archives)

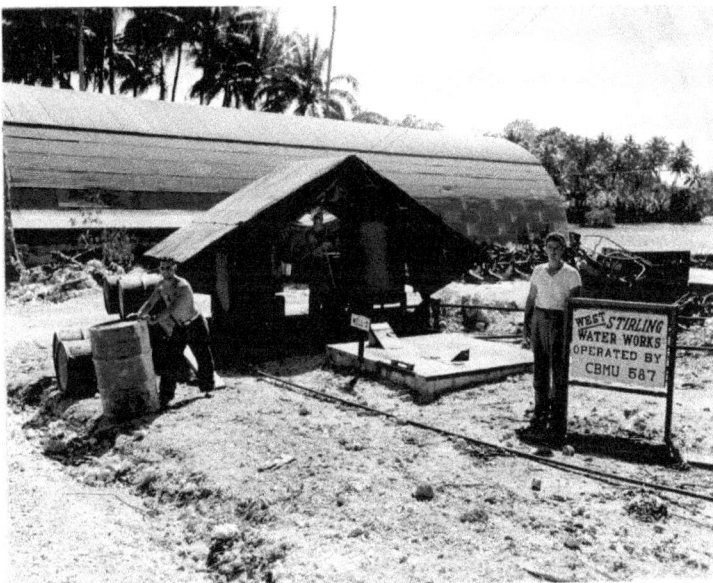

Water well number two at the base was operated by the Seabee Maintenance Unit 587. This unit arrived at the base on May 3, 1944. Clean drinkable water was paramount for all those at the base. (National Archives)

Catholic Mass services held at the base for both Sailors and Seabees. Conducting the services is Lt. Patrick Donohoe (USNR) of the 82nd Construction Battalion. (National Archives)

9. Lone PT Boat heading into the base at Treasury Island. This wonderful photo shows the effect of the wake, caused by the boat's powerful engines and hull design. From the air, even at night Japanese planes could spot the boats from a sizable distance. (National Archives)

10. Comedian Jack Benny performs at the base as part of their Pacific USO tour. The tour featured singers Martha Tilton and Carol Landis along with Harmonica Player Larry Adler, and pianist June Bruner. They ended the tour on September 15, 1944. (Lt. (jg) Ross E. Anderson Jr.)

Sailors at the base take time to pose with the pretty girls of the USO Jack Benny tour. From left is Carol Landis, Martha Tilton, and June Bruner. From the looks on their faces, they are a happy bunch of young men. (Lt. (jg) Ross E Anderson Jr.)

Wonderful port side look at Higgins *PT-237* as the she heads back to Treasury Island base. She would serve with Squadrons 19 and 20 in the South Pacific before being transferred to the Southwest Pacific area. (Lt. (jg) Robert Davis Helsby)

A look at the dry dock area, showing *PT-155*. Regular maintenance was crucial to the boats' upkeep, and these dry docks gave the crews and base forces the ability to check the boats bottom, props, shafts, and struts. They could also do proper upkeep of the boats' mufflers and the scrapping and painting of the hulls. (Frank J. Andruss Sr.)

A tracked vehicle called a tractor crane, sometimes referred as a crawler crane was used to supply the boats with their torpedoes. Here a MK-XIII torpedo is being hauled to the waiting roll off rack on the boat. (Frank J. Andruss Sr.)

Boat captains at the base have gathered to take a group photo. Behind them sailors are struggling to carry something, while a PT Boat can be seen in the background. (Lt. (jg) Ross E. Anderson Jr.)

This Landing Ship, Tank (LST) has nosed its way onto the beach to unload much-needed supplies. These wonderful ships were developed to support amphibious operations and could carry tanks, vehicles, cargo, and troops. As can be seen in this photo, these ships did not need docks or piers to land their cargo. (R. W. Spencer)

Cape Torokina

The invasion of Puruata Island took place on November 1, 1943. It took place on the same day as the main Allied invasion of nearby Bougainville and saw a force of Marine raiders capture this small island close to the main American beachhead. The main American landings took place around Cape Torokina on Empress Augusta Bay (on the western coast of Bougainville). Puruata Island is about half a mile from this beachhead and was garrisoned by a platoon of Japanese infantry. It was to be attacked during the first wave of the American invasion. The attack was to be carried out by the 3rd Raider Battalion (Lt. Colonel Fred D. Beans), with one reinforced company in the lead and the rest of the battalion as a reserve.

The landing was opposed by light fire, and by 9.30 the Marines had established a secure perimeter around 125 yards deep. They were facing snipers, machine guns and mortars, and so at 1.30 p.m. the rest of the battalion joined the attack, supported by some self-propelled 75mm guns. The battalion then launched an attack that saw them occupy half of the island by the end of 1 November. On 2 November the Marines launched a two-pronged attack on the Japanese half of the island. This time they only faced rifle fire, and by 3.30 p.m. the island was secure.

Twenty-nine Japanese bodies were found, and the rest of the garrison escaped to Bougainville. On the morning of November 3, 1943, US Navy (USN) Commander Henry Farrow arrived with eight PT boats to establish a PT boat base on Puruata Island, known as Torokina PT Boat Base (Torokina Base) due to its proximity to Torokina. The PT boat base and boat pool were set up on Puruata Island by the 75th Battalion, assisted by the 71st and the seventy-seventh Battalions. Wood-pile and timber construction was used for a PT boat pier, a crash boat pier, and a PT fueling pier.

Complete camp facilities included quarters, mess halls, an emergency hospital, with all utilities, and five prefabricated steel warehouses. Eighteen small-boat moorings, consisting of three pile dolphins, driven and lashed, were provided, and LST landings installed. Stevedoring was handled by the 6th and the 9th CB Battalions. PT boats from this base covered the western coast of Bougainville and as far north as Buka.

November 1, 1943: Landing Craft, Vehicle, Personnel (LCVPs) from the USS President Jackson (APA-18) off the beach are screened by PT boats as they continue to Empress Augusta Bay on the Western end of Bougainville. This operation would be known as "Cherry Blossom" and were the amphibious landings by the 3rd Marine Division. They were met with limited resistance and by November 3rd they were firmly established, when Torokina Island was occupied. (National Archives)

PT boat entrance sign to the base. Painted on the sign are the words "Give me a fast ship for I intend to go in harm's way.", said by American captain John Paul Jones on November 16, 1778. (National Archives)

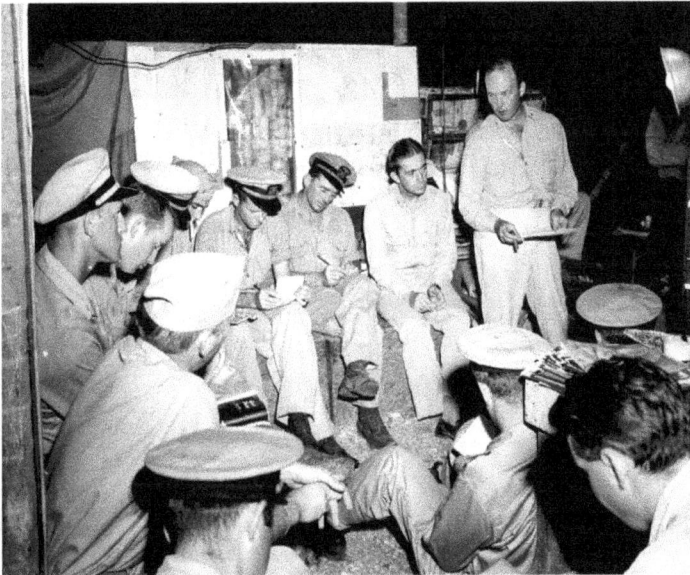

Boat captains are crowded into this small operations tent as they listen to Cmdr. Thomas G. Warfield. He is the commanding officer of the base and is briefing them just prior to the boats leaving for patrol. (National Archives)

Interior look at the bakery on the base. Here fresh bread would be baked for all PT crews to enjoy. Ovens were made from gasoline drums, brick, and mortar. The baker in the photo is baker 2/c H.M. Garathorp. (National Archives)

A look at the tent area on the base that made up the repair shops. In this area contains the electrical and machine shops. Many components were electrical in nature on a PT Boat so it was instrumental that the base had men that could take care of all electrical needs for the boats. Machine shops could take care of many needs from welding to making new parts. (National Archives)

Here in the electrical shop, Momm2/c G.A. Jones is working on an auxiliary generator for one of the boats. These generators were instrumental on the boats in making sure components and equipment in the chart-house had steady and reliable current. Most important to the crew was these generators also ran the refrigerator, cook top, and coffee maker. Generators were used when the boats engines were not running. (National Archives)

A look inside the metal-smith and ship-fitting shop at the base. On the left W. B. Bradley SF2/c is using an acetylene torch to cut steel, while O. G. White Momm2/c and R. C. Steen CM3/c perform other duties. (National Archives)

PT-254, a Higgins PT boat, is getting her 40mm mount replaced. The crane was instrumental for this type of heavy lifting, which saved considerable time and manpower at the base. This crane was a typical ten-ton machine and used for PT boat service. Many times they were mounted on a 6-by-18-pontoon barge. (Frank J. Andruss Sr.)

Inside the electrical shop H. R. Klatt EM2/c (left) is repairing a broken terminal while H. C. Greer S1/c takes hydrometer readings on the PT boats batteries. Keeping batteries in tip top shape was yet another system on the boats that required the utmost attention (National Archives)

A look inside the armory and ordinance shop, showing (left) M. Lyttle S1/c checking a.30 caliber carbine rifle while N. N. Colbert CM2/c is working on one of the boats 20mm cannons. All weapons on a PT boat had to be in 100 percent working order before that boat left for a patrol. (National Archives)

While docked at the base, this Higgins PT boat crew is enjoying a meal prepared by the boats cook. The Higgins boat had their galley as part of the boats crews' quarters as can be seen in this photo. The large ladder in the center leads up to the boats chart-house and topside. (National Archives)

Some of the base force staff crowd around this truck to take time for a photo. The base force performed specialized work after going to school for instruction in the construction, maintenance, and repair of the PT boats. They were instrumental in the continued upkeep of not only the boats but other machinery that was on the base and performed a variety of work. (National Archives)

PT-167 docked, serving with RON 10. The boat was attacked November 6,1943 by Japanese torpedo planes as both the PT boat and LCI(G) 70 were retiring from Cape Torokina to the Treasury Islands, Solomon's and were both struck by dud torpedoes. The hit on the 167 would pass clean through the bow, leaving an entrance and exit hole. (Frank J. Andruss Sr.)

This is the Base operations tent that served as the office for the Base Captain, Executive Officers, and Intelligence officer at the PT boat base. (National Archives)

This is the underground area of the sick bay on the base. Above this area was the First Aid tent. Below ground they could attend to patients where they had several beds. An operating table with adequate lighting was available as well as all types of medicines and other medical equipment. In the case of a bombing raid, patients would be safe here. (National Archives)

This is the men's living quarters at the PT boat base. The tent structure in the center of the photo is raised over a bomb shelter. In the foreground is another log structure in case of any bombing raids. (National Archives)

PT-175 and her crew take time for a photo. Notice two of the crew are wearing open toed sandals, while others wear an assortment of clothing, from dungarees with two crewmen wearing overalls. Two of the crew are sporting beards, while one has the words "Handy" on his helmet. (Bob Hart MoMM1/c)

Here mechanics are repairing 6-71, two stroke HN-9 Gray-Marine variant engines, probably for LCVPs (Landing Craft Vehicle, Personnel) or Higgins boats. Normally inspection, service, and repair would be conducted inside repair shops at the base, but noisy conditions, poor ventilation, and space concerns made it ideal for working outside. (US Navy Seabee Museum)

Sailors bring a bit of home to the island as they play Bingo inside one of the recreation tents. Bingo was and still is a popular game that can be played for cash and prizes. During the conflict, bingo and a wide range of raffles and lotteries became a principal means of raising funds for war-related charities and other causes. (US Navy Seabee Museum)

Green Island

Moving up from the Torokina PT Base, the occupation of Green Island would add a new Allied base for offensive sweeps well beyond the previous range for South Pacific aircraft. Green atoll consisted of four, flat, thickly wooden islands which almost encircled a lagoon. This area would provide for not only aircraft but for the little PT boats as well. Construction would begin on the airfield and shortly after gasoline tank farms were set up, road, communications, and PT boat pier service line with all the connections. By June of 1944, the Seabees would construct a fuel pier with a 35-foot outboard end with the PT base being developed to include a camp, four shops with approach ramps, one prefabricated steel warehouse and a pontoon type T shaped pier with a 154-foot outboard end.

Medical facilities constructed would consist of floored and screened tents as well as a Quonset hut that housed X-Ray facilities. They also erected four Quonset huts for the Navy Base Hospital. PT boats would be moored at the Green Island Lagoon. While establishing the PT boat base at Green Island, PT tender (AGP-5) USS *Varuna* was now situated at the base perhaps nearest to Japanese territory.

Located halfway between New Ireland and Bougainville, both occupied by the Japanese and both well north of the nearest Allied bases— Green Island served as a staging area for the five MTB squadrons attached to Varuna. These PTs were earmarked to harass the Japanese seaborne supply lane from New Britain, New Ireland, and the Shortlands, and to assist in the blockade of Rabaul. *Varuna* operated out of Green Island until 31 July, when she returned to the Treasury Islands for a 20-day stay. Returning to Green, she loaded men and equipment of Motor Torpedo Boat Squadron 27.

One aspect of the Allied victory in the Pacific was the speed with which newly captured islands were set up. On the Green Islands the PT base was operational by 17 February, only two days after the start of the attack. C.J.

Willis who served on *PT-242* and was stationed at Green said: "We operated quite often with Black Cats (PBY Catalina planes) on our patrols out of Green. They would drop flares in Japanese coves so we could see if there might be barges etc. in there. Also they told us if they saw something that they wanted the boats to investigate."

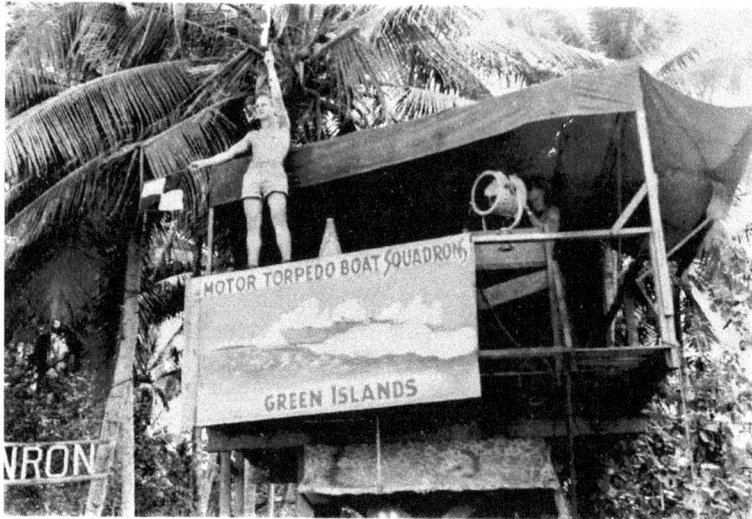

Entrance to the PT boat base at Green Island. Over the sign is a camouflaged signal tower. In this photo a sailor is using semaphore flags to signal incoming boats. (Frank J. Andruss Sr.)

This small duty boat was kept busy at the base, as side trips to other small islands would provide much-needed information. The duty boat had a variety of uses that included mail runs, fishing trips, and the all-important trip to larger ships nearby to try and get some better food for the crews. (C. J. Willis)

LST-39 at the beach to off load much-needed equipment and supplies at the base. Notice the unknown PT boat that is nestled near the ship's stern. *LST-39* was laid down on April 23, 1943, by the Dravo Corp. at Pittsburgh, Pa.; launched on July 29,1943; sponsored by Mrs. L. A. Mertz; and commissioned on September 8,1943. She was assigned to the Pacific area during World War II but saw no combat action. She sank in the summer of 1944, at West Loch, Pearl Harbor on May 21,1944 when mortar ammunition that was being loaded aboard LST -353 exploded. Later raised and used as a barge. She was struck from the Navy list on July 18,1944. (R.W. Spencer)

A look at the PT tender USS *Varuna* (AGP-5) anchored just off the island. The ship tended the boats as support while the Green Island Base was being completed. This ship when fully loaded, displaced 377 tons and was 328 feet long. *Varuna* would become one of the first Allied warships to enter Manila Bay after the Japanese surrender of Corregidor. (National Archives)

With the jungle somewhat cleared out, the buildings built here include the Carpenter shop (left) and the engine overhaul shop. The truck parked here was used to transport supplies and other parts from the dock to these buildings. Base force personnel were kept busy making sure the boats were operational for every patrol. (C. J. Willis)

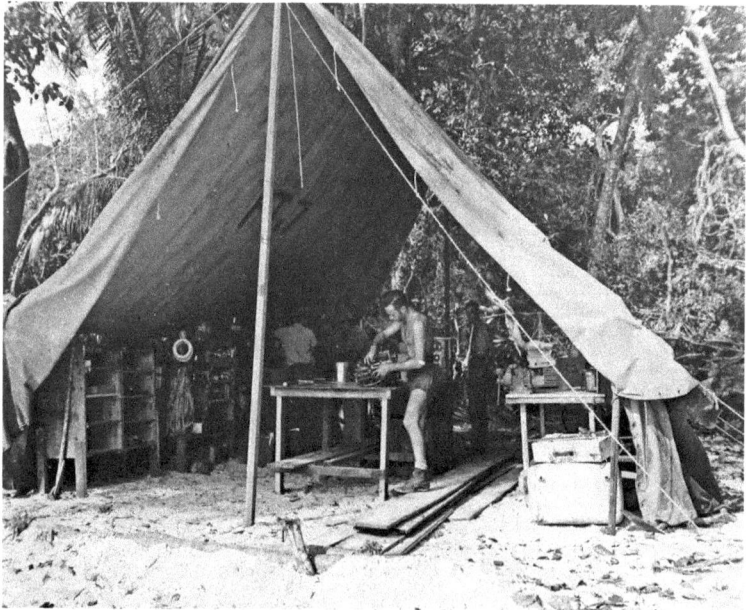

A simple but effective repair tent set up at the base. Here Holly carburetors are being fixed, adjusted, and cleaned. When completed they will be installed on the Packard marine engines ready for testing. Many times a shortage of supplies made it difficult to keep the boats engines running smooth. At times parts from other boats would be used to supply those boats going out on patrol. (C. J. Willis)

A look at the typical officers' quarters at the base. Notice they have been built using a tent structure on wooden walls. They have screened in windows and awnings to keep the rain out. Bugs were a constant problem, so screens were a much-needed addition to aid in the comfort of the men. Malaria was always a problem, so these quarters were built to deal with mosquitoes. (R.W. Spencer)

Nicknamed "The Greasy Spoon" this was the ship's store. This Quonset style building provided items such as Candy, writing paper, and Tobacco products. Prior to the Seabees building these structures, the base was nothing more than leaky tents with boats moored to bushes and floats. The identity of this officer is unknown. (C. J. Willis)

This area was cleared out by the Seabees and would be used for Mass Services and entertainment. Logs were constructed that would act as seating, with movies shown here early on once a week, then every night as movies were shipped in. Here Sunday services are taking place. (C. J. Willis)

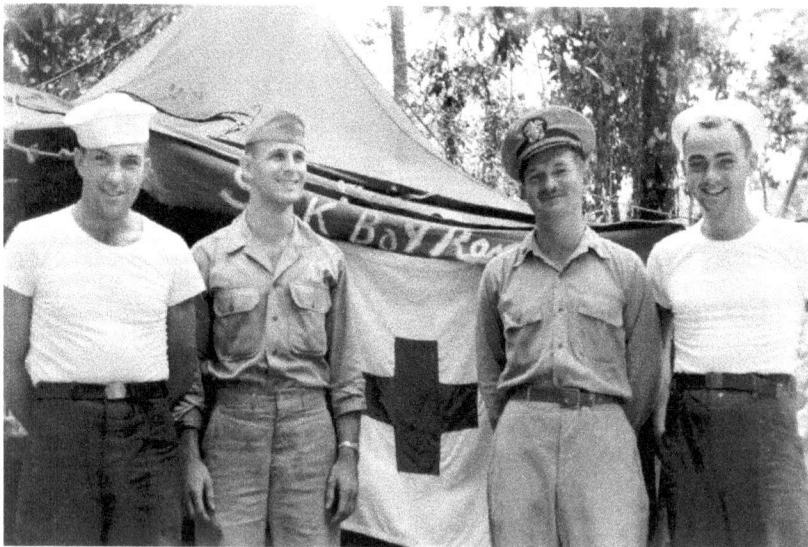

The medical staff on the base stand in front of their tent. Supplies for the department were usually shipped on time, although delays sometimes occurred. All medical emergencies were treated here, with the more serious cases being moved to more advanced medical facilities. (C. J. Willis)

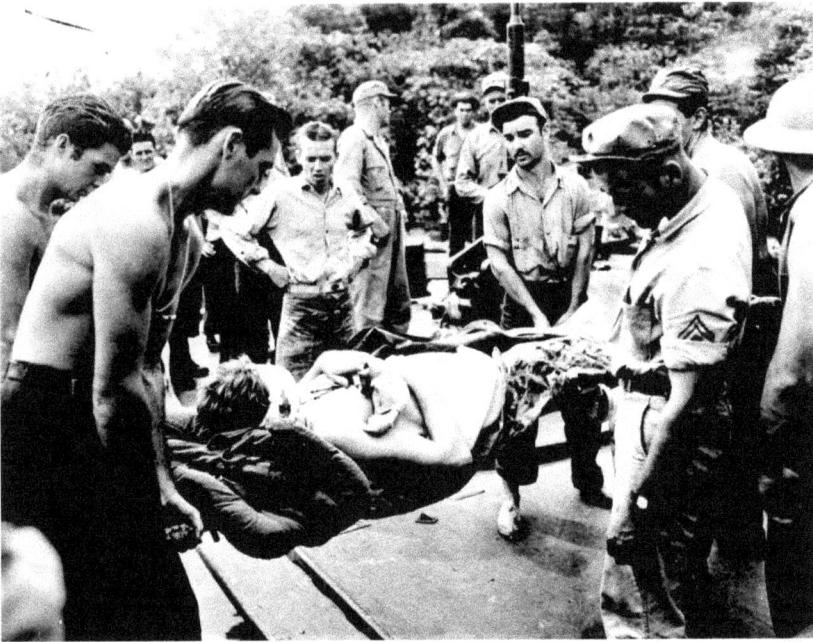

Wounded Sailor is being stretchered from this PT Boat. His injuries appear to be of a serious nature, and he will be evaluated by the medical staff at the base. If need be, he will be removed to the advanced Hospital off base. (National Archives)

PT Boat officers gather outside of the officers' quarters to sit and relax. They seem to be enjoying a much-needed smoke. They will talk about patrols and families, probably telling each other what they will do once they get home. Later, things will become somber and serious as they prepare for the nights patrol. (C. J. Willis)

An officer from this unknown Higgins PT boat has certainly dressed the part. With hard helmet and a Sub Thompson machine gun, he is ready for patrol. He is also carrying the standard web belt with a.45 caliber side pistol and a two-clip magazine pouch. Side arms like this were usually carried by the boat crew. (C.J. Willis)

The outdoor Bob Hope USO Tour show at the base. Prior to the show, the sailors and others had been waiting since lunch for them to get to the base and start the show. Sitting on those coconut logs was tough on the backside. As things at the base were pretty secure no one had to stand guard duty. When it started to rain singer Francis Langford stopped and asked those in the crowd, "What do you do when it rains?" The crowd of PT Sailors yelled "We get wet." (C. J. Willis)

Officers checking out the 37-mm auto cannon on this boat. Early on the M-4 cannon was mounted on numerous US Navy PT boats as deck guns, starting with the Solomon Islands campaign. Primary targets were landing barges used by enemy forces to move supplies down the island chain at night. Initially, they were taken out of crashed P-39 aircraft at Henderson Field. When they proved successful as anti-barge weapons, they were used in this role for the rest of the war. The 30-round endless belt magazine was given the designation M6; it had an oval-shaped framework (nicknamed a "horse-collar magazine" from its shape) providing a track for the endless belt. Beginning in 1944, the M9 cannon was installed at the builders' boatyard as standard equipment. (Frank J. Andruss Sr.)

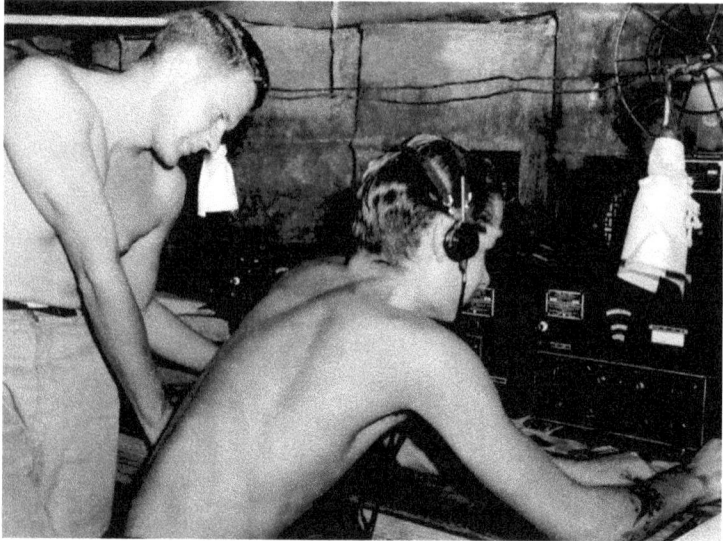

Radio communications shack at the base. These men are listening in on any Japanese movements in the area. Reports will filter in, and be given to the squadron commander, who in turn will get them the to the boat captains. A briefing usually follows, and selected boats will be given their night assignments for patrol. (C. J. Willis)

A look at the all-important dry docks that were used at the base. Instrumental throughout the Pacific as the little boats would enter the docks and be raised out of the water for hull maintenance. Marine growth in the Pacific would attach itself to the boats hulls and slow the boats speed. In the dry docks, crews would scrape and paint the boats bottom and sides. Rudders, props, and shafts as well as mufflers would be worked on here. (Frank J. Andruss Sr.)

This Higgins PT boat, probably from Squadron 23, pulls into the Green Island base. The squadron was under the command of Lt. Comdr. Ronald K. Irving, USN, later being transferred to the Southwest Pacific area. (C.J. Willis)

PT boat operating around the Green Island Base. Crew are gathered around the stern 20-mm Oerlikon gun. The weapon would carry a Spiral magazine which held sixty rounds and were spring driven. To aid in spotting, a tracer was usually included in every fifth round. By the end of 1944, the 20-mm Oerlikon's importance had been overshadowed by the 40-mm Bofors' greater hitting power and longer range. Notice the midships 20mm has no magazine attached. (C. J. Willis)

EMIRAU

Moving up the line another PT base was established on Emirau Island located at Hamburg Bay on the Northwest of Emirau Island. Also known as PT Boat Base Emirau Island (Hamburg Bay). The boats were supported by PT Boat tender USS Mobjack (AGP-7) operated from Homestead Lagoon on the west of Emirau. Emirau Island PT Boat Base was used by the US Navy in 1944. The US Navy Seabees of the 18th Construction Regiment arrived between the 25th and 30th March 1944 and commenced building the PT boat base, a floating drydock and slipway, and roads, another section built the ammunition storage facilities, a runway, and some of the buildings of the PT boat base.

Operating from this base was part of General MacArthur's plan to encircle the Japanese base at Rabaul. The landing on Emirau marked that final step in MacArthur's plans to defeat the Japanese in the Pacific Theater.

The island was completely undeveloped before World War II, with a single track along its length, together with a spur into the northern peninsula and a maze of trails in its western portion, and the Japanese presence was limited to a few coast watchers, but even these had been withdrawn by January 1944. PT boats here would operate patrols off New Ireland.

The dock set up at Emirau can be seen in this photograph. Dock space was available for the boats at both sides, with a crane in the center for heavy lifting of engines or torpedoes. At the closest end to the beach is a dry dock for maintenance work, and as one can see this area was compact and efficient. The base was located on what was known as Hamburg Bay, on the Northwest of Emirau Island. (National Archives)

The torpedo shop area shows torpedoes lined up and ready, as needed for placement on the boats. The Quonset hut was the main area of the shop where all repairs and other work would be performed. Notice to the left the tracked torpedo hauler, called a crawler crane that could easily haul torpedoes from this area to the waiting boats at the dock. (Frank J. Andruss Sr.)

Local villagers on Emirau gather in their home. The local language was a dialect of the Mussau Emira. More than 300 natives and their families occupied this Island. (National Archives)

Taking time out for a beer, a popular pastime with sailors. From the number of empty bottles in the case at the center, they are enjoying it very much. The beer they are drinking is Acme Beer who promoted patriotism back home as they actively encouraged numerous means to aid the war effort. They advocated giving blood; planting Victory gardens; writing to the troops; recycling cooking grease to your butcher; and other economizing activities. (Frank J. Andruss Sr.)

A post card looking photograph as this PT Boat is captured though one of the trees on the island. If not for the war this photo would have one believe that this spot is a wonderful place for a vacation. (Frank J. Andruss Sr.)

Some of the crew of *PT-180* take time for a quick photo while at the base on Emirau. The boat would serve with Squadron Eleven under the command of Lt. Cmdr. Leroy T. Taylor, USN. (Frank J. Andruss Sr.)

PT-65 heading into the base. Crew gather around the bridge of this Elco seventy-seven-foot boat in 1944. The boat would serve with Squadron 5. After assignment to the Southwest Pacific area at the end of 1944, the squadron was decommissioned, and its boats were distributed to replenish other squadrons which had suffered operational losses (Navsource photo)

Some of the crew of *PT-176* gather on the bow of the boat. The sailor third from the left in the back row is providing music while playing the accordion. This boat would serve with Squadron 11. Prior to this photo the boat was damaged by Japanese float-planes on August 24, 1943, while patrolling at Gizo Strait, Solomon's. (Bob Hart MoMM1/c)

Sailors Goforth (left) and Paul Ivanics take time for a quick photo. Both sailors are crew members of *PT-176*. Sailors in the South Pacific dressed to combat the heat and humidity. Some preferred no shirt while others wore the typical US Navy "chambray" shirts worn by sailors throughout WWII in all theaters. While on patrol most wore shirts. (Bob Hart MoMM1/c)

Paul Ivanics and Alongi leaning up against the day-room cabin on *PT-176*. To the left of Alongi you can see the box structure that is holding the semaphore flags rolled up. (Bob Hart MoMM1/c)

Bob Hart MoMM1/c (back row left) and his buddies have managed to secure some grass skirts. It was not uncommon for PT boat sailors to enjoy a bit of fun by dancing and clowning around for crew members on the boat. Grass skirts were usually made from natural beach grass and were hand knotted. Sailors many times would send them home as sweetheart gifts for a wife or girlfriend. (Bob Hart MoMM1/c)

PT boat gasoline tanks by the roadside, having been replaced. The wing tanks (there were two) for the Elco PT boat held 850 gallons each while the center tank held 1300 gallons. These were self-sealing tanks, although some of the boats did not get them until after their engines were replaced. PT boat squadrons consumed hundreds of gallons of high-octane fuel daily. (Robert Douglas GM)

Index Of Persons

INDEX OF SHIPS AND BOATS

Index of Technologies

The dazzling range of technologies identified here were required to support multiple squadrons of PT boats, tenders, and their crews at rugged, isolated bases scattered across the vast Pacific. A useful corrective for naval enthusiasts who may focus too tightly on the movement of ships on maps.

PT-170 RON 10 escorting ship 1943. (National Achieves via Ted Walther).

www.ingramcontent.com/pod-product-compliance
Lightning Source LLC
Chambersburg PA
CBHW050404110426
42812CB00006BA/1794